CLASSIC PLANT MACHINERY

CLASSIC PLANT MACHINERY

BRIAN JOHNSON

BOXTREE

In Association with
Channel Four Television Corporation

3 8015 01309 1420

Acknowledgements

This book is based on the Channel 4 television series, 'Classic Plant and Machinery'. A great deal of original research was accomplished in the preparation of the series by the producer, Uden Associates. I was in the fortunate position of having unrestricted access to that research in the writing of this book, and of receiving the personal assistance of the producers and researchers of the television programmes: Michael Proudfoot, James Castle, Senara Wilson, Gemma Chapman, Anna Whelpdale, Ellis Gwilym and Sophie Williams. I also acknowledge with gratitude the invaluable help I received from my patient editor at Boxtree, Katy Carrington, and the skill of designer Robert Updegraff in achieving such a result in the short time available to us.

Brian Johnson,
CATTISTOCK, DORSET

First published in Great Britain in 1998 by Boxtree

an imprint of Macmillan Publishers Ltd
25 Eccleston Place, London, SW1W 9NF
and Basingstoke

Associated companies throughout the world

ISBN 0 7522 2437 9

1 3 5 7 9 10 8 6 4 2

A CIP catalogue record for this book is available from the British Library.

Designed by Robert Updegraff
Printed and bound in Italy by New Interlitho

Half title page: *The first Aveling & Porter steamroller of 1865 (Road Roller Association Archive Collection).*
Title page: *An operator's-eye view from the cab of a dockyard crane (Tony Stone Images/Christian Lagereek).*

Classic Plant Machinery accompanies the Channel 4 series
'Classic Plant ' produced by Uden Associates.

Contents

AO-385

CHAPTER ONE

The Tractor: A Horse for All Seasons

Tractor *One who or that which draws or pulls something; esp. a traction engine 1856.*

THUS DOES THE *Shorter Oxford English Dictionary* dismiss a technology that was to be the means of the greatest change in British agriculture since the Enclosure Acts, which from 1761 progressively enclosed, for private use, over 5 million acres (2 million ha) of common land. From time immemorial this had been freely available for cultivation by yeoman farmers.

Opposite This Ivel tractor was built as a works demonstrator in 1901. It was sold in 1903 to a Mr Tinneswood and has remained in private hands ever since. The tall chimney was to clear any steam from the cooling tank and stop it from obscuring the driver's view. Apart from the old car-tyre tread wrapped round the front wheel and another tyre of sorts lashed to the rear wheels, the machine is very original and remains in working order.

Farming is basically a physical activity. Farmers – or the men they employ – spend their time either putting things into the ground or digging them out. They prepare the earth by ploughing, sowing, harvesting and dragging heavy wagons as dictated by the four seasons. In the earliest times, when life was brutal and short, this work was performed by human muscle. One of the first representations of humans performing as tractors, though not in an agricultural context, was made as long ago as 2000 BC in Egypt (*seen below*). It is a bas-relief depicting no fewer than 172 slaves dragging an alabaster statue 23ft (7m) high on a simple wooden skid. The strength of 172 men is the nominal equivalent of 21 horses; even so, in the light of a recent televised experiment, a similar number of fit young students had the greatest difficulty in moving a concrete replica of a Stonehenge stone column on level, greased skids. On that evidence, even 172 men would seem a marginal number; perhaps each figure in the relief represented 10 men or the artist simply filled up the cartouche. Whatever the true figure, the ancient relief illustrates that when it comes to shifting heavy loads, human effort, even expendable slaves in unlimited numbers, are a puny source of power.

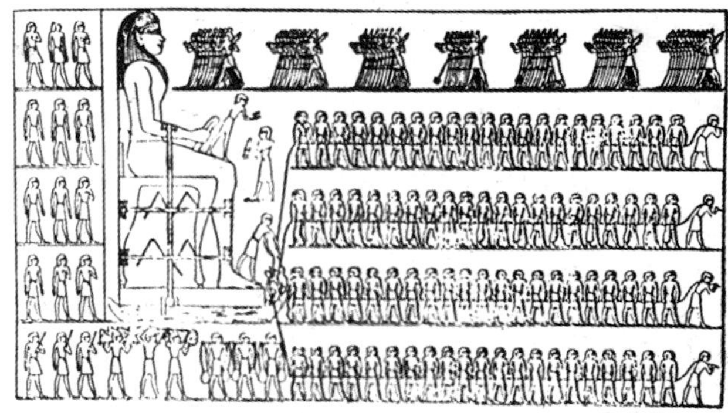

Above In Ancient Egypt, when slaves were plentiful and expendable, they were prime movers even if it took 172 to haul a statue. Recent experiments make it doubtful if even that number could drag the weight of the statue depicted very far.

A fit man can perform work in horsepower hours equivalent to one-eighth of that of which a horse is capable. But in antiquity even the horse was inhibited in its power output by the simple yoke harness, with body and neck girths, which, when a heavy load – a plough for example – was drawn, caused the windpipe and ribcage to be depressed. This restricted the animal's breathing and therefore the work it could perform.

The horse collar was a simple invention that, with the wheeled plough, revolutionized agriculture from around the 11th century and remained the standard method of ploughing until the arrival of the motor tractor in the 20th century. (Woodcut from: *Virgil*, Opera. Strassburg: Grüninger 1502.)

However, the yoke harness suited the slow-moving ox (in the Third World it still does) but not the much faster and more powerful draught-horse.

If the Enclosure Acts and the replacement of the horse by mechanical horsepower are the twin pillars of modern agricultural development, there is also a third, but earlier, consideration – the appearance, around the tenth century, of the collar-harness which allowed a horse to haul much heavier loads due to unrestrained breathing. One of the earliest illustrations of the 'new' harness is to be seen in the Bayeux tapestry: the scene depicts a mule ploughing with the old girth and a horse drawing a harrow with a collar-harness.

The increase in traction of which the horse was capable led to another significant development: the heavy-wheeled plough which two horses could now draw over even difficult ground. There is some doubt as to when the wheeled plough was first put to use, but it is thought to be around the eleventh century. There is a woodcut from Virgil's *Opera*, published in Strasbourg in 1502, which clearly depicts both the heavy plough and the collar-harness in everyday use. The horse-drawn heavy plough was to permit greatly increased agricultural production and was, with little fundamental change, to remain the basic method of ploughing in the Western world for 500 years.

One of the problems of the heavy plough was the necessity of keeping the relatively soft cast-iron share sharp but because of a happy accident in the foundry of the Ipswich plough-maker, Robert Ransome, a very tough share was produced by a process known as 'chilled iron'. The hard share cast in chilled iron, retained its edge and was fitted to the firm's famous iron-beam ploughs, which were introduced in 1843. It was to prove to be the classic British plough and many were made until production of Ransome's horse-drawn plough ceased in 1948.

In the heyday of horse-ploughing, a team of two good Suffolk Punch horses led by a skilled ploughman – or horseman, to give him his correct title – could plough, on light land, an acre of standard British 9-in (22.5-cm) furrows in a working day; a day, incidentally, which began for man and horse at 4am. On heavy soil, typical of most British counties, three-quarters of an acre (0.3ha) was the acceptable norm. In achieving this, man and horse would have walked over 12 miles (18km) of newly turned furrow. At the turn of the century the ploughman would have been paid for his hard-won skills a maximum wage of 15 shillings (75p) a week, though he would have had a rent-free cottage and perhaps wood for fuel.

At the rate of three-quarters of an acre (03.ha) a day, it follows that a typical 5-acre (2-ha) meadow would take a single team a working week to plough it. With the exception of the four years of the First World War (1914–18) this was considered satisfactory, and horse-drawn ploughs were a common rural sight right up to the Second World War of 1939. A team of horses drawing a plough could still be seen, though in dwindling numbers, into the 1950s and a few, in the hands of enthusiastic owners (some of whose ploughs can be up to a hundred years old) linger on to this day.

These ploughs are reminders of a vanished world – a world that was governed by the slow-changing seasons. The fields were small with layered hedgerows shaded by tall elms;

Ploughmen, like this one photographed taking part in a traditional ploughing match in 1938, were true countrymen. Their way of life and work had changed little over the years, and they regarded the tractor as an unnecessary, noisy intrusion to the traditional horse-drawn plough and the gentle rhythm of the changing seasons.

the horses were immaculately turned out, with braided manes and tails and glittering brasses on the martingales. They worked with soft-spoken encouragement from the ploughman, who skilfully guided both horse and plough. The harness clinked with the soft singing of the share as another, ruler-straight, furrow of dark earth was added to the stretch.

Of course it could never last. The first signs of change were the steam railways that began in 1825 between Stockton and Darlington and, within a decade, spread across the countryside, no doubt frightening the horses as they did so. However, steam power, the mainspring of the Industrial Revolution, made little impact on agriculture. Steam traction-engines appeared in the middle of the nineteenth century, the first being built in 1841 by the same Ipswich engineering firm, Ransome, which made the ploughs. These magnificent engines, fully developed by the turn of the century, were, as we shall later see, to be used for threshing, clearing woods and as a static power source for heavy sawing.

Although traction-engines were used in the United States directly to pull giant ploughs turning up to twelve furrows at a time, what could be done on the level, treeless prairies of the Midwest was not possible in the cramped, eccentric, small fields which, in Britain, were the legacy of a thousand years of complex land tenure. The small size of the undulating average British field or meadow also made the double steam-ploughing engines, popular both in Europe and the United States, have limited appeal.

The Fowler Company of Leeds was the first to have patented 'ploughing engines' in 1856. These machines, which were an adaptation of the standard traction-engine of the time, worked in pairs, standing opposite each other on the headlands of the field to be ploughed. Each self-powered engine of 12–14nhp had a large winch drum beneath the boiler driven by bevel gears and a dog clutch under the control of the driver. A steel hawser, typically 450yd (411m) long (half a mile was not unknown), was paid out through sheaves at right angles to the engine and

A delightful line drawing from the pages of the magazine *The Engineer*. The engine is a 1858 Fowler 14hp twin-cylinder ploughing engine. This is a 'left-hand' machine as the cable for the plough leaves the sheave on the left-hand side. The man from *The Engineer* was critical: 'The whole tackle requires unceasing attention which it is not likely to get from the ordinary farm labourer…'

A Fowler ploughing engine no.15347, built in 1919 and still in working order in 1997. It differs remarkably little from the 1858 design. This one, too, is 'left-handed'. Originally these engines would be supplied as a set: one left, the other right-handed. They would then both face the same way as the furrows advanced across the land being ploughed.

connected to a wheeled, centre-pivot reversible plough, which typically had five shares at each end. Since both engines faced the same direction, the engines were 'handed' – the engine on the left of the field was right-handed (i.e. the wire left the drum on the right-hand side), the right-hand engine being opposite to take up the draw wire on the left.

The plough, looking like a giant, 30-ft (9-m), iron-lattice seesaw, was balanced on two 5-ft (1.5-m) wheels. The end that was away from the drawing engine was pulled down on to the small guide-wheels at the end of the seesaw. The ploughman ensured that the five active shares were set at the correct depth for ploughing. When set, the near engine-driver would give a toot on his steam whistle and then, with an answering whistle from the distant engine, the massive 30-ft (9-m) plough would be majestically drawn across the field. The ploughman, perched high on the plough, steered the 5-ft (1.5-m) centre wheels with a huge steering-wheel, keeping the furrows straight – always a point of great pride for any ploughman, whatever the means of drawing his plough. When the plough arrived at the end of its travel, whichever engine was drawing, the shares were lifted clear of the furrows. Both engines advanced the width of the turned stretch, usually 6ft (1.8m), and then the first engine drew the reversible plough back.

The 'set' consisted of the two steam engines, the plough or cultivator and any ancillary implements. A five-man crew was normal: a foreman, whose main function was to negotiate with farmers, two engine-drivers, a ploughman plus a cookboy. All, both men and machines, were hired from contractors.

A steam-ploughing crew, dependent on demand and season, had to be prepared to live like gypsies. They worked from farm to farm, dawn to dusk, twelve to sixteen hours a day for weeks on end. Their weekly wages, as late as the 1930s, would have been £10, divided by the foreman between the five-man crew. It follows that from the men's point of view, the most vital part of the steam-ploughing set was the 'living van' in which the

Steam ploughing in full flight with the driver keeping the furrows straight. This photograph was taken around 1940 when steam power staged a comeback due to the shortage of petrol for tractors during the war years. It was a short-lived reprieve. The heyday of the ploughing engine had passed thirty years before.

crews ate and slept. The living van was towed from field to field by the engines, along with the plough or cultivator, water wagon and a supply of good Welsh steam coal for the engines and living-van stove.

A good steam-ploughing team could, under ideal conditions, plough a 40-acre (16-ha) field in a 16-hour working day. However, conditions were seldom ideal and the average acreage tilled was a good deal lower. In the first place, the field to be ploughed had to be level, otherwise the plough shares would burrow out of sight; the field had to be of reasonable size and shape to make the expense worthwhile; the enterprise was, as so often is the case in agriculture, dependent on the weather; and the soil of the field to be ploughed had to be dry and consolidated sufficiently to support the 10- to 20-ton deadweight of each engine without undue compaction. That limitation, of course, also applied to the access route to the field. Finally, due to the size of the engines, the headlands (those portions of the field between the end of the furrows and the boundary hedge) was considerable and had to be ploughed in by a conventional horse-drawn plough after the engines had departed.

Steam-ploughing was invariably put out to contract since only the very largest estates would have cared to pay the £2,000 that the engines and ancillary equipment would have cost around 1900 (£200,000 in today's money). One of the main steam-ploughing contractors before the First World War, Ward & Dale, had twenty-five steam-ploughing sets available for hire from their Sleaford, Lincolnshire depot. The Dorchester-based contractors, Eddisons, served the needs of farmers in the West Country.

Apart from ploughing, the contractors also offered mole drainage, dredging of lakes and ponds and the cutting of drainage ditches. Even so, the steam-ploughing engines, magnificent though they were, found little favour in Britain. In general, the fields were just too small and irregular. The transit times from the contractors' depots to the farms must have been on the long side too as, until 1903, steam tractors, which were required by law to have a two-man crew when en route, were restricted to a maximum speed of 2mph (3km/h) in towns or villages (this was so as not to frighten the horses). Once on the open road they could gather speed up to a heady 4mph (6.4km/h). After the Heavy Motor Cars Order of 1903, a small (5-ton) engine was permitted to steam, with a single driver, at 5mph (8km/h).

These swingeing restrictions did little to encourage a more widespread application of the steam plough in Britain. They were more successful elsewhere, particularly in Germany and the United States. However, although the Fowler Company remained the main builders of these specialized machines, four or five other traction-engine builders (Burrell of Thetford, Lincolnshire, and Aveling & Porter of Rochester, Kent, being the best known) still listed ploughing engines in the 1920s – mainly, one suspects, for the export market.

At the peak of the steam plough's popularity, during the intensive drive to increase British agricultural cultivation during the First World War, over 600 sets were in use. A surprising number survive. About 160 sets, mostly in working order, delight people at summer steam fairs; a few intrepid owners even offer their services for contract work, usually dredging rivers and lakes, though an occasional ploughing contract is still possible. Long may they continue to steam!

Although steam-ploughing engines were the first serious attempt at agricultural mechanization and were used on large estates, principally on the flat, wide land of Lincolnshire, they did not, in general, even begin to threaten traditional horse-drawn methods. At the turn of the century, it is unlikely that the traction-engine builders, working with inextinguishable Victorian confidence in their smoke-filled, thudding foundries and erection shops, ever gave a passing thought to the notion that the new petrol-driven 'motor cars' which had just appeared could even begin to challenge their fiery monsters as prime 'movers'. They were proved seriously mistaken.

The first 'motor tractors' appeared within four years after the motor car had demonstrated that the petrol-fuelled internal-combustion-engined vehicle was a practical proposition. The word 'tractor' is believed to have been coined for an early petrol-driven agricultural machine made by the Hart-Parr Company of Iowa, USA, in 1902. Little is now known of that pioneer effort; indeed, there would be many false dawns before the classic and definitive farm tractor, the American Fordson Model F, first appeared in 1917.

Many of the early American names are now little more than footnotes to agricultural history: Burger, Froelich, Van Dusen, and many ephemeral devices, most of which existed only as unsatisfactory prototypes or design proposals. The first British design to make it into actual metal was one of the earliest. It was the grandly named Hornsby-Akroyd Patent Safety Oil Traction Engine, which appeared in 1896. Unfortunately that ambitious pioneer effort suffered a serious weight problem and as far as is known only one example, probably the prototype, was ever sold.

The first successful British-built agricultural tractor is without doubt the Ivel Agricultural Motor designed by Daniel Albone, who was a racing cyclist and constructor

Daniel Albone, the father of the motor tractor.

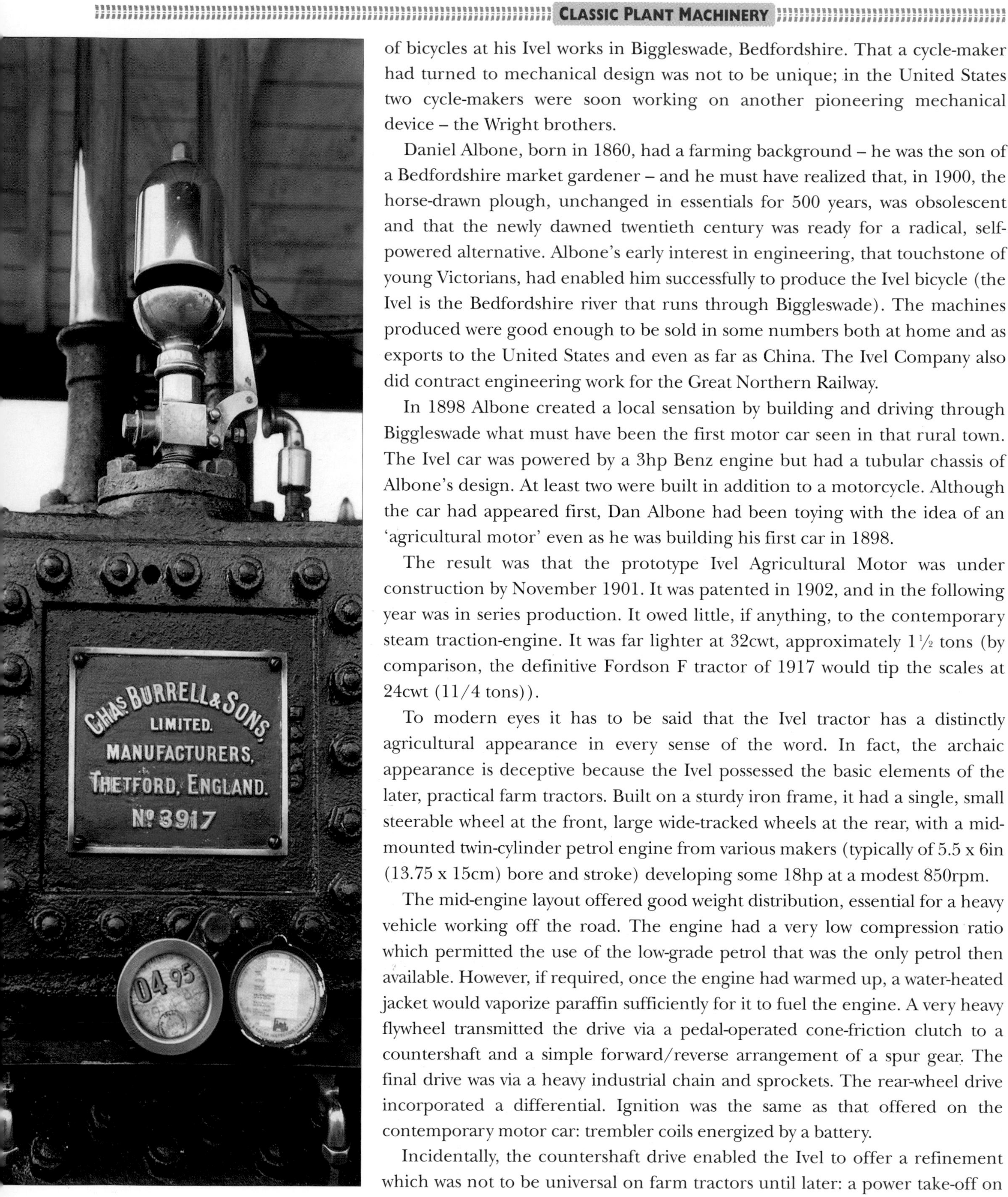

of bicycles at his Ivel works in Biggleswade, Bedfordshire. That a cycle-maker had turned to mechanical design was not to be unique; in the United States two cycle-makers were soon working on another pioneering mechanical device – the Wright brothers.

Daniel Albone, born in 1860, had a farming background – he was the son of a Bedfordshire market gardener – and he must have realized that, in 1900, the horse-drawn plough, unchanged in essentials for 500 years, was obsolescent and that the newly dawned twentieth century was ready for a radical, self-powered alternative. Albone's early interest in engineering, that touchstone of young Victorians, had enabled him successfully to produce the Ivel bicycle (the Ivel is the Bedfordshire river that runs through Biggleswade). The machines produced were good enough to be sold in some numbers both at home and as exports to the United States and even as far as China. The Ivel Company also did contract engineering work for the Great Northern Railway.

In 1898 Albone created a local sensation by building and driving through Biggleswade what must have been the first motor car seen in that rural town. The Ivel car was powered by a 3hp Benz engine but had a tubular chassis of Albone's design. At least two were built in addition to a motorcycle. Although the car had appeared first, Dan Albone had been toying with the idea of an 'agricultural motor' even as he was building his first car in 1898.

The result was that the prototype Ivel Agricultural Motor was under construction by November 1901. It was patented in 1902, and in the following year was in series production. It owed little, if anything, to the contemporary steam traction-engine. It was far lighter at 32cwt, approximately 1½ tons (by comparison, the definitive Fordson F tractor of 1917 would tip the scales at 24cwt (11/4 tons)).

To modern eyes it has to be said that the Ivel tractor has a distinctly agricultural appearance in every sense of the word. In fact, the archaic appearance is deceptive because the Ivel possessed the basic elements of the later, practical farm tractors. Built on a sturdy iron frame, it had a single, small steerable wheel at the front, large wide-tracked wheels at the rear, with a mid-mounted twin-cylinder petrol engine from various makers (typically of 5.5 x 6in (13.75 x 15cm) bore and stroke) developing some 18hp at a modest 850rpm.

The mid-engine layout offered good weight distribution, essential for a heavy vehicle working off the road. The engine had a very low compression ratio which permitted the use of the low-grade petrol that was the only petrol then available. However, if required, once the engine had warmed up, a water-heated jacket would vaporize paraffin sufficiently for it to fuel the engine. A very heavy flywheel transmitted the drive via a pedal-operated cone-friction clutch to a countershaft and a simple forward/reverse arrangement of a spur gear. The final drive was via a heavy industrial chain and sprockets. The rear-wheel drive incorporated a differential. Ignition was the same as that offered on the contemporary motor car: trembler coils energized by a battery.

Incidentally, the countershaft drive enabled the Ivel to offer a refinement which was not to be universal on farm tractors until later: a power take-off on the left-hand side that could drive, via a standard belt, a threshing machine or a saw bench. Though a useful addition, the use of the Ivel as a stationary engine was not its primary role; heavy ploughing and field cultivation were the main functions.

The intrepid driver sat high on the right-hand side, slightly behind the rear wheels. His high position permitted him an excellent view of the ground immediately ahead, an important point when ploughing. The wide steering-wheel was horizontal and indicated the cycle-maker/designer in that it was connected to the front-wheel fork by a bicycle-chain. A firm hand was required as, unlike modern steering via non-reversible gears, the inevitable kickback from the unsprung front wheel had to be constantly countered if the tractor were to maintain an accurate course, essential for good ploughing. To the left of the driver sat a very large water-tank containing no fewer than 30gal (136l) of cooling water. The large quantity of water was necessary as the Ivel lacked a conventional radiator and the additional weight over the rear wheels was considered an asset in increasing adhesion. Be that as it may, the hot tank sitting alongside must have been a great solace to a driver working a frozen field on a cold morning in early spring.

Contemporary photographs exist of the Ivel being demonstrated with Daniel Albone in proud attendance. It is interesting to see that the machine is ploughing with

Opposite The brass whistle and safety-valve covers of a Burrell Traction Engine no. 3917 is one of a number of magnificent machines of the steam age which is still preserved in working order today. It was getting on for 80 years old when this photograph was taken at the 1997 Blandford Steam Fair.

Below This historic photograph was taken c.1902 during one of the early field demonstrations of the prototype 'Ivel' tractor at which Daniel Albone himself was present with his son (*front right*). The land wheel of the Ransome's horse-drawn plough can be seen just behind the wheel of the tractor.

standard Ransome beam horse plough, necessitating a second man to handle the plough. That 50 per cent increase in labour costs, even at 15 shillings (75p) a week, could have been one of many reasons why the Ivel was to be a moderate commercial success. Later, towards the end of production, the Ivel Company offered a rather daring remote control which enabled the driver to sit on the implement being towed, thus reducing the crew to a single man.

The 1903 photograph, possibly taken at a farm at Beaulieu, is of one of the earliest public trials of the Ivel. The rather grand location was a consequence of the support Daniel Albone enjoyed from the Hon. John Scott Montagu and S. F. Edge. Edge had raced cycles with Albone in their youth, and he was now the creator and driver of the Napier racing car. Another name connected with the Ivel was the racing motorist Charles Jarrot, who had driven a Panhard et Levassor in the heroic age of motor-racing. All three of these well-known sporting Edwardians were fellow directors with Albone in the formation of Ivel Agricultural Motors Ltd, formed in 1903. Beaulieu apart, Albone had concluded an agreement with a local Bedfordshire farmer, a Mrs Kendall, to have the use of a field to test, develop and demonstrate the Ivel.

The press must have been invited to witness some of the demonstrations, for the London *Daily Mail* was to report that: '. . . farmers stared in astonishment at an Ivel motor at work yesterday . . . as it ploughed a double furrow through hard, frost-bound land, in which eight horses could only plough one furrow.' *The Times* thundered that: '. . . harvesting in Lincolnshire has been conducted with the aid of a petrol motor specially invented for farm use. The motor . . . cut a field of barley at Tinwell . . . the motor has also been tried with a double furrow plough, and cut some deep and even furrows through

BROTHERS IN ADVERSITY

Farmer: 'Pull up, you fool! The mare's bolting!'
Motorist: 'So's the car!'

This 1901 cartoon from *Punch* is typical of the many published at that time at the expense of the 'motorist' who was a popular butt of the day. It reflected, however, a genuinely felt aversion – particularly among countrymen – to machines in general, and one which was to inhibit the adoption of the agricultural tractor on British farms for far too long.

stubble.' *The Country Gentleman* made the prescient observation that: '. . . the machine, when it can be placed on the market in sufficient numbers, is likely to be used even more in the colonies and abroad than in this country.' *The Engineer* reported that 'The [Ivel] machine . . . should meet with a considerable demand.' And *The Car* prematurely enquired: 'Is the Horse Indispensable?'

The notices were good and the farmers quoted by the *Daily Mail* may well have been astonished. However, although Albone tried, with some enterprise, to promote his tractor, few farmers, at least in Britain, seemed to have reached for their wallets. Sales of the pioneer agricultural machine were very sluggish.

Sadly, Daniel Albone died in 1906 at the early age of 46, leaving his agricultural motor in a relatively underdeveloped state. Despite gold medals won in Italy and at the 1904 Royal Show, the Ivel failed to gain acceptance. Even as late as 1910, when the Ivel had several competitors, tractors in general were still lacking support down on the farm. In 1910 the Royal Agricultural Society, in a commendable effort to encourage mechanization, arranged a trial of 'Tractors' with a gold medal for the winner. Seven machines, including a late Ivel model and three steam tractors, competed in the trials. The results were depressing for the motor-tractor makers. After four months of deliberation by the judges the result was no outright winner, the judges declaring that: 'Steam engines best fulfilled the requirements of the trials.' They did, however, add a rider to the effect that: '. . . the oil [petrol] engine would ultimately best suit the farmer's requirements if a general-purpose motor were to be adopted for the average farm.'

The Ivel, alas, was not to be that ideal machine. As John Frankenheimer said of one of his films: 'It went from rave reviews to classic without ever being a success . . .' This is a sadly fitting epitaph for the Ivel – the Ivel Agricultural Motor company went into receivership in 1916. Two Ivels are known still to exist: one is in private ownership and the other, splendidly restored, is in the London Science Museum – a fitting tribute to engineer Daniel Albone.

Why was the Ivel, certainly the most practical tractor of the time, unable to secure a viable home market? The answer is complex. In the first place it should be remembered that the easy synergy we now have with cars and machines generally did not exist in 1903 when the Ivel went into production. Most farmers had never driven a motor car; possibly many had not even seen one. They were, especially in the country, perceived as noisy, dangerous devices liable to run out of control and very possibly explode, and also as the playthings of rich, irresponsible young men. That apart, people regarded petrol with the same degree of apprehension that we would regard a can of nitroglycerine. The horse lobby, which was large and vociferous, played to those fears and received unquestioning acceptance from traditionalist farmers.

Farmers had to be traditionalists; they were, and still are, trading in the future; 'futures' are subject to risks from nature and the folly of man. The margins are slender: a failed crop is a lost crop and it can never be retrieved. The main advantage claimed by the tractor men was speed. But the clock of farming is set by the slow rotation of the four seasons. A crop raised from 25 acres (10ha) of ploughland will be the same as if the field had been ploughed by one man and two horses in a week or by one – or two – men and an 'agricultural motor' in a morning. There was, moreover, no risk with the horses. Horse-ploughing had been refined over 500 years and was, as we might now say, 'fully sorted'. The new tractor, with its explosive fuel and temperamental engine, most definitely was not. Even if the Ivel had been proved to be completely safe and reliable, the most persuasive fact against it was the price: the Ivel cost £450 in 1906. That was, at that time, 'serious' money. With £450 in 1906, a farmer could pay a farm-hand his wages for twenty years or he could buy his son his first, modest farm.

The final demise of the Ivel in 1916 was inevitable as by then it was out of date. With Daniel Albone gone, the life went out of the firm and the tractor was not developed in any significant form. It was doubly unfortunate as at that time, ironically, British agriculture was in a crisis so deep that the government had to act.

The problem was simply that, due to increasing urbanization following the Industrial Revolution, and unlike France which had retained a self-supporting peasant-based agrarian economy, Britain could no longer feed herself and had to import food. As early as 1887, the country was importing 700,000 tons of meat and 19 million tons of grain annually. With the development of refrigeration ships, from a tentative start in 1877, the figures of imported meat increased dramatically. By 1912 there were 218 refrigerated ships on the British register, importing nearly 16 million carcasses from New Zealand and Argentina. The figures for imported grain had also risen greatly.

In peacetime such importation is a matter of economics but when, in February 1917, the German U-boats began a ruthless, unrestricted sinking of merchant ships the situation deteriorated rapidly into crisis. Churchill was to write of April 1917: 'In April the great approach route to the south-west of Ireland was becoming a veritable cemetery of British shipping.' In that dire month a total of 516,000 tons of British shipping and 336,000 tons of Allied vessels had been sunk by German U-boats; an unsustainable rate of loss, far beyond the capacity of British shipyards to replace.

The British government faced, at best, a reduction of the civil food ration to a point where vital war production could be compromised or, at worst, being forced, by the threat of starvation of the civil population, to sue for peace – in other words, to surrender. There were two alternatives. One, bitterly contested by the Admiralty but in the end the most successful, was that of putting ships into convoys, protected by naval vessels. The second was the belated realization by the government that, following the run-down of British agriculture caused by the importation of cheap food from the Empire, the country possessed a large amount of fallow land that was suitable for arable use.

There was to be a drive to cultivate those unused fields for cereals and potatoes, but there was a grave difficulty. The manpower available for work on the land had decreased as a consequence of the war. Many young farm workers had joined the forces, some out of a genuine patriotic conviction egged on by jingoistic music-hall songs sung by large contraltos such as 'We don't want to lose you, but we think you ought to go'. Others had been shamed into going off to war by young women who placed the white feather of cowardice in their hands. However, 'voluntary' enlistment was, by 1916, insufficient to replace the tens of thousands of young, ill-trained soldiers dying daily in the mud of the Western Front. By the end of 1915, the numbers of volunteers were becoming insufficient to meet the insatiable demand of the forces for men. In January 1916, the Military Service Act introduced compulsory conscription. So a new seam of manpower was opened. Hundreds of thousands of men went to join the war, many of them farm workers, and their horses were also commandeered.

Military historians write that the First World War was the first mechanized war. This is only true when applied to tanks, aircraft and, where the few good roads allowed, heavy lorries. Overwhelmingly, the armies of the fighting powers relied on men and horses to drag heavy guns through the mud of the battlefields, to bring up the ammunition, the rations, and to remove the dead and the wounded and bring forward the replacements. Horses were needed for the cavalry and for the generals posing for the photographers and the new cinema newsreels – horses that might have been drawing the plough.

The military remount depots worked around the clock breaking horses from Ireland and the United States as well as from Britain. Horses are pathetically vulnerable to small-arms fire and shrapnel; the losses were appalling. Horses were being killed by the million, and the demand for replacements was endless. In a normal peacetime year British farms alone required around 50,000 foals to replace worn-out workhorses. In 1917, nothing like that figure was available for civil use. Even if the horses had been available, the loss to the countryside of the skilled horsemen, farriers, blacksmiths and so on, could not be replaced; it simply took too long to acquire the skills. The thousands of young volunteers to the Women's Land Army working on the farms could never be trained in sufficient numbers to plough with horses in time to surmount the crisis.

The solution was to use tractors. The newly recruited women of the Land Army could rapidly be trained to drive and plough with a tractor. Moreover, a tractor could work as long as there was daylight – for twenty-four hours if given artificial lights. In contrast, the horses drawing a plough traditionally had to be returned to stables by 2.30pm; if they were worked beyond that the law of diminishing returns followed. But where to find the 7,000 tractors that made up the first target? British makers were either bankrupt like Ivel, or had tractors of unsuitable design or were simply unable, due to other essential war work, to produce anything like the numbers so desperately needed.

There was, however, another source: the United States. Henry Ford, the man who had invented the mass-production line at his factories in Dearborn, Michigan, and had produced the immortal Model T car, which was capable itself of a little light harrowing, was the necessary man. A farmer's son himself, Ford had long wished to produce a reliable farm tractor which would be cheaply produced in great numbers – a Model T of the land.

The Ford company had been experimenting with prototype tractors from as early as 1907. It was not until 1915 that Henry Ford had set up a company, Henry Ford & Son Ltd, as a separate concern from his car production company. It had been allocated some of Dearborn's best engineers; they were, in effect, given a clean sheet of paper and told to invent the farm tractor. No doubt Henry Ford's philosophy, which was 'simplify and add lightness', though originally applied to motor-car design, was still ringing in their ears.

The team was led by Eugene Farkas, who is credited with the idea of dispensing with a conventional chassis by making the engine, gearbox and rear-axle castings of sufficient strength to accept the stresses of the whole as a monocoque single unit. It was a brilliant concept, saving weight and both production time and cost.

It might be whispered that this idea was not, perhaps, totally original, but the execution of the Ford design far surpassed an earlier attempt at unit construction made by the Wallis tractor company. However, the Wallis Cub of 1913 had a U-shaped undertray made of boiler plate, which enclosed and supported the engine, gearbox and transmission into a neat package. It was not a genuine monocoque; simply a U-shaped chassis.

Once the Ford design was finalized a small number of pre-production machines were built to test the tractor in the field. It was immediately obvious that the 'Fordson' was a winner: it was a purposeful four-square design with a low centre of gravity, which was essential when working across hilly ground. The tractor's innovative unit construction ensured that oil would stay in the engine, and that dust and mud, inseparable from cultivation work, were kept out.

The wide-tracked front-axle beam was pivoted at the centre from a strong attachment point, which was part of the massive crankcase casting. The sturdy, relatively large, cast-iron front wheels, each with ten flat spokes, had a raised flange along the centre line of the rims to bite into soft ground and keep the tractor heading straight. Full car-type non-reversible ackerman steering made the driver's task easy. No brakes or springing were fitted to the front wheels.

Above the axle attachment, the crankcase casting also supported a large conventional multi-tube radiator, the top of which also located the oval fuel tank that fed the engine by gravity. The other end of the fuel tank was mounted on a somewhat vestigial bulkhead which in turn supported the steering-wheel, hand throttle and ignition switch. The rear wheels, which were about a third as large as the front, were substantial spoked castings with wide treads, each having sixteen diagonal strakes across as an aid to traction and to lessen wheelspin. Only later in the production run would rear mudguards be fitted. Pneumatic tyres were never an option to the Fordson Model F during its entire eleven-year production life.

The driver sat on the tractor rather than in it, astride the gearbox casing. There was no floor, so his feet sat on footrests, similar to the steps still to be seen on telephone

The definitive agricultural tractor. This Fordson Model 'F' shows the salient features which made it a world leader from its introduction in 1917. The iron-shod wheels, without any form of springing, offered good traction on soft ground but must have given the driver a rough ride on a hard surface. Pneumatic tyres were never an option. The vestigial footrest can be seen just to the front of the rear wheel. The captive starting handle was the only way of starting the engine.

poles, bolted to either side of the gearbox casing. The driver did have a perforated cast-iron saddle seat moulded to what Ford considered to be average, ample agricultural buttocks. The seat was supported above the differential casing by a short strip of flat steel which acted as an unyielding spring – the only springing, incidentally, on the tractor. One wonders what it was like driving a Fordson on a frozen lane or hard-surfaced road, even with a pile of empty seed sacks which acted as the usual cushion.

The engine of the Fordson Model F, open to the elements, with only the fuel tank as a cover was, throughout the eleven-year production run, to remain unchanged in essentials. It was a very substantial water-cooled, four-cylinder, side-valve unit, known in the United States as a flat head. A side-valve engine, while lacking the elegant mechanical efficiency of an overhead valve of equivalent size, is simple and has far fewer moving parts and thus will retain its admittedly moderate tune for a long time. The compression ratio was low, indeed low enough for canny British farmers to discover that, once hot, the engine would happily run on paraffin (kerosene). Later, Fordsons had a double fuel tank which could be switched from petrol for starting and warming up to paraffin in order to run. Later still the oil companies offered a superior fuel to paraffin lamp oil, known as Tractor Vaporizing Oil (TVO). However, in its native country with, in 1916, gasoline costing only about 5 cents a gallon (4.5l), the kerosene/paraffin option was not required.

The Fordson's four cylinders had a bore and stroke of 4 x 5in (10 x 12.5cm) producing 22hp at 1,100rpm. The Model F in working order weighed in at a very modest 1 ton 4cwt (1.2 tonnes). This, in 1916, was far lighter than any conceivable British rival, some of which tipped the scales at three times that weight. The turning circle, a very important point when working small fields, was 22ft (6.5m). The ignition system on the early production was the same coil/battery and trembler as was used on the Ford Model T. Later a magneto was fitted, but never a self-starter. Starting was by a captive handle which, with the low compression, was not difficult. However, the driver did have to remember to retard the ignition before cranking if he or she did not wish either to be thrown across the farmyard or suffer a broken wrist. Once running, it goes without saying that there was no protection for the driver against the elements. No doubt the early drivers consoled themselves that in that respect they were no worse off than they had been when ploughing with horses, and that now they did not have to walk.

The above description of the Fordson Model F is, undoubtedly, that of the most influential farm tractor yet produced. It was the definitive tractor; all the others that were to follow, right up to the present day, have been developments. Such was the impact of the Fordson Model F that, once in production, most other tractors became obsolete and many rivals simply disappeared.

This is anticipating events, however. The position in 1916 was that the British government was desperate for very substantial numbers of an improved tractor. It has to be said that the government was very fortunate. At the hour of need in 1916 a small number of pre-production pilot model Fordson Fs had arrived in Britain from the then still neutral United States, sent by Ford to test the British market. They were seen widely at ploughing matches and country fairs which, perhaps surprisingly, still took place despite the war.

Ford, far from having to indulge in a 'hard sell', found that he was pushing against an open door. As soon as reports of the performance of the new tractor were received, the

Looking at the Saunderson Universal Model 'G' of 1916, it is hard to believe that it was a near contemporary with the Fordson 'F'. In comparison, it was much heavier and cumbersome and failed to appeal either to farmers or to the British government who, in 1917, ordered a large number of the American tractor instead.

government invited Ford to set up a factory in Britain to build the Model F. An initial order for 7,000 units was guaranteed. Ford declined mainly on the grounds that the increasing activities of the German heavy bombers, the infamous 'Gothas', were causing real apprehension when, in June 1917, they raided London in broad daylight. The public outcry was such that the government had to announce a very big increase in the production of fighters and anti-aircraft guns to defend the capital and country. Ford possibly took the view that the supply of raw materials and the skilled workers they needed would be diverted to the munitions factories; in any case, from April 1917, the United States was also at war with Germany and its industry was put on a war footing.

As an alternative to a British factory, Ford agreed to set up a tractor production line at Dearborn. Incredibly, by October 1917, just four months after the British order, the first of the 7,000 units were being mass-produced and shipped across the Atlantic in the holds of American ships. The new grey-painted tractors with the distinctive bark of their exhausts were soon to be a familiar sight and sound on British farms. It is a measure of the urgency of the situation, and the contribution made by farmers to the war effort, that the distribution was through the Ministry of Munitions.

To the farmers who were fortunate to get their hands on one of the Fordsons, they came as a revelation. The tractor was light, very reliable and simple to drive even when drawing a traditional beam plough; hastily trained land girls and boys too young for the army could plough 10 acres (4ha) in a day; the furrows might not have been ruler-straight but the fallow set-aside land, much of it uneconomic or too difficult for horse-ploughing, was put under cultivation. Ploughing apart, the Fordson was also to permit a markedly increased efficiency in the other agricultural tasks of harrowing, seed-drilling and mowing, while still using the standard horse-drawn implements.

The use of thousands of tractors, together with the effectiveness of the convoying of merchant ships, which reduced shipping losses due to German U-boat attacks, narrowly avoided the threatened severe shortage of food. By the end of 1918 the tables had been turned: the Allied naval blockade of Germany had caused incipient starvation to force Germany to sue for peace. On 11 November 1918, the First World War, after four blood-soaked years, ended.

After the euphoria that the armistice brought the surviving farm workers, demobilized, returned to the land. The Fordson tractors, bought by the government, and which had been seen as a possible salvation in 1917, were sold off and continued to work on the farms of those farmers who were lucky enough to buy them at knock-down prices. Tractors had clearly become part of the rural scene. British tractor-makers anticipated a growing market. It was not to be. The British government, under the

imperative of war, had in effect invited the American Fordson company to corner the tractor market to the exclusion of emerging British manufacturers. Henry Ford & Son Ltd were well aware of the situation and lost no time in establishing a tractor factory at Cork in Ireland in 1919. The factory was, by 1921, turning out the first postwar Model F tractors, listed at £205 in England. The market was to be dominated by the Fordson; by 1922 the price was reduced to £120 in a ruthless move by Ford to undercut all possible competition (ironically, much the same situation, though this time with transport aircraft, would arise twenty years later at the conclusion of the Second World War).

The British Index of 'Agricultural tractors 1921–1929', prepared for insurance companies, lists no fewer than thirty-one makers of farm tractors. By 1929 most have 'No Particulars' or 'Discontinued' recorded against their entries. Only two names of the thirty-one quoted in the 1921–9 Index, Case and Ford, remain in the tractor business today. Significantly, both are American. The only British tractor that might have challenged the Fordson, which was the Saunderson, is listed in 1922 at £330, nearly three times the price asked for a Fordson.

It was not only in Britain that Ford tractors were driving the competition out. The same situation occurred in the United States for the same reason: Ford could afford to slash prices because of their cost-effective design, resources and expertise in mass-production techniques. A small number of US makers, however, did survive the Ford near-monopoly. The J. I. Case Threshing Machine Company of Racine, Wisconsin, which had started out as steam traction-engine makers, imported into Britain several models of Case tractors in the postwar period, ranging from 10 to 30hp, and priced from £235 to £395, the amount asked by the English importer, Associated Motors of King's Cross, London.

Perhaps the most popular Case tractors were the C & L models of 18–32hp introduced in 1929, which impressed with their performance during the grandly named 'World Tractor Trials' held at Benson, Oxfordshire, in 1930, in which practically all the available makers were represented. The Case machines were deemed to justify their sales slogan 'Do not consider price alone', possibly because they were far more expensive at £395 than the ubiquitous Fordson which, in fact, broke down during the trials.

Another American tractor that sold in Britain in the early thirties was the orange-painted Allis-Chalmers. This large 40hp tractor offered a power take-off and was a popular alternative to the steam threshing engine of the time. However, the Allis Model U was an expensive proposition at over £460. The company is best remembered as one of the first, perhaps actually the first, to offer the option, in 1932, of wheels shod with the newly produced Firestone pneumatic tractor tyres. These had deep treads that offered the same grip as the traditional cleated metal wheels without damage to the newly appearing tarmac country roads. It is possible that it was to gain publicity for the new tyres that the Allis Company prepared a specially geared Model U to establish an unofficial speed record for a farm tractor at a staggering 68mph (109km/h). Other tractors of the 1930s that survived the Fordson sales drive were the various models made by the International Harvester Company, another American firm which, like Ford, opened a British factory, this one being in Bradford, Yorkshire.

Fordson (as the company was generally known) continued producing the Model F up to 1929. From 1922 the tractor was produced in the Ford factory at Trafford Park, Manchester. By the time the eleven-year production of the Model F ceased, nearly three-quarters of a million had been produced worldwide. Fordson enjoyed 75 per cent of the market, so it is small wonder that the competition had a hard time. The successor to the Model F was the 28hp Model N. Introduced in 1929, it bore a close family likeness to the earlier Fordson but it had put on weight, at 2 tons 3cwt (2.2 tonnes), double that of the Model F. The Irish factory was to produce the first 38,000 between 1929 and 1932, when production was switched to the newly opened Ford plant at Dagenham, Essex.

Whereas the Model F swept the board because it was inexpensive and at the same time fulfilled its design objects very well – as well or better than far more expensive rivals – the Fordson Model N was to face severe and challenging opposition. Briefly, the new Fordson had the same simple side-valve petrol engine, though now with magneto ignition, which removed the necessity of a battery. Following the lead made by Allis-Chalmers, Fordson, from 1932 onwards, offered the option of Firestone pneumatic tyres. The new Fordson also had mudguards to protect the driver, relatively speaking, from the mud flung from the rear wheels.

By the time the Fordson N was in production, colour had begun to be important. Allis-Chalmers (as noted) were orange – not ordinary orange but 'Persian Orange'.

Later Case machines were 'Flambeau Red' or 'Prairie Gold'. The Fordsons, during the sixteen-year production, were finished originally in blue; by 1938 this was changed to orange (shade unspecified) and, finally, in 1940 to green. In 1940, British Fordsons produced in Dagenham were camouflaged. The RAF used thousands to tow both aircraft about the airfields and the trains of bomb trollies laden with high explosives from the bomb dumps to the waiting bombers. Jim Wilkie of Old Sodbury owns a restored wartime Fordson N: 'This tractor literally stopped Britain from starving to death and saved countless lives at sea as it reduced the North Atlantic convoys. I always think it symbolizes Anglo-American unity.'

From 1933 the Fordson Model N, despite enjoying a major share of the world market, was obsolescent. It lacked hydraulics and was, in effect, an improved version of the 1917 Model F: a drawer of implements. It survived because it was a good, rugged, very reliable general-purpose tractor and it still enjoyed Ford's trump card – it was cheap. It cost £155 when introduced in 1929 and never exceeded £180 during its sixteen-year production run. On average, the Fordson N was almost half the cost of any comparable rival. But cost is not everything. In 1933 there appeared a single prototype British tractor which, though retaining the classic outline as defined by the early Fordson, was to define the technology of a future generation.

The new tractor looked at first glance much the same as the contemporary Fordson N. It appeared rather squatter, an impression gained from the long and very wide-tracked wheelbase. The tractor had the, by then, usual unit construction, though it was painted in an unfashionable sinister jet black. The only white paint on the machine was the maker's name in an elegant script on the top of the flat radiator. It was a single word: 'Ferguson'.

Harry Ferguson was born in 1884 on a small family farm in County Down, now in Northern Ireland. He soon displayed a natural aptitude for engineering, together with a fertile imagination and the ability of lateral thinking. He left the family farm and, with his brother, ran a garage. He found the time to design, build and pilot his own aircraft which became, in 1909, the first to fly in Ireland. Not content with that landmark, in 1911 Ferguson acquired the agency for Vauxhall motor cars, which he also raced with some success.

The garage also had the agency for an early, heavy American tractor with the improbable name of 'Waterloo Boy', which Harry Ferguson demonstrated without much commercial success to local farmers. (The successors to the Waterloo Gasoline Engine Company still exist, trading under the name of John Deere.)

Like Daniel Albone and Henry Ford, Harry Ferguson, coming as he did from a farming background, had worked the land and knew the strengths and the limitations of contemporary tractors. He realized that, by the early 1930s, the classic farm tractor, the Fordson N, had reached a plateau of development. As an innovative engineer, Ferguson came to the conclusion that an updated Fordson, a tractor designed just to tow its implements, however well, would never be able to compete with the American giant. But Harry Ferguson decided that there was a vacant niche, overlooked by the men from Detroit which, he decided, he could profitably exploit even against the mighty international Ford organization. Harry Ferguson had foreseen something that Ford and others had not: that the next advance in agricultural mechanization lay not in a better tractor but in a fundamental reappraisal of the role of the tractor and of the essentially

This Amercan tractor, the Waterloo Boy of 1916, had the distinction of having Harry Ferguson as the UK agent, before he had even thought of designing his own tractor. However, even he could not sell the expensive machines. The company still exists, now trading under the respected name of John Deere.

unchanged horse-drawn implements, the ploughs in particular, which it towed.

Some changes had been made. The classic horse-drawn trailer plough with its two asymmetric wheels, the small land wheel and the larger adjustable depth wheel, guided by the ploughman with the two handles at the rear, had been supplanted around 1919 by a plough designed for tractor work. This was the Ransome RSLD, or self-lifting plough. The RSLD had two equal wheels but the shares could be set for depth of plough, or raised clear when at the end of the furrow, by forward-facing controls which the tractor driver could engage manually. He could do this by simply turning around and operating cranks, while keeping a hand on the steering-wheel. This type of plough, however, was still trailed behind the tractor, connected via a simple flexible drawbar, and this method was to prove to have a very serious shortcoming.

When ploughing virgin land, because tractors – typically the Fordson – were both powerful and lightweight, if the shares struck an immovable object below ground (a tree stump, for example) the tractor, with most of the weight on the rear wheels for a good grip, would rear up like a motorcycle performing a 'wheelie'. This remorseless application of Newton's third law of motion was so abrupt that the driver seldom had the time to declutch or close the throttle, the result being that he, if lucky, was thrown clear; if not, the somersaulting tractor crushed its luckless driver. As early as 1922, this had been the cause of no fewer than 136 fatal tractor accidents involving Fordsons in the USA. All were caused by the plough striking an immovable object, and the tractor then rearing up and crushing its driver.

Such accidents were fortunately rare in Britain as most fields had been ploughed for generations; buried tree stumps and rock ledges had been discovered by horse ploughs and removed long before. However, the weight distribution of contemporary tractors was such that if a driver let in the clutch suddenly with the plough or heavy implement attached, even though the machine did not have sufficient momentum to somersault the tractor would lift the front wheels clear of the ground and the driver would lose control. The same situation applied if, during ploughing, a patch of exceptionally heavy earth was encountered. The Fordson N had the rear mudguards extended to 'tool boxes' which were, in reality, there to act as stoppers in the event of the machine rearing up. In practice, they possibly did no more than give the driver a little more time to declutch.

The black Ferguson entirely removed the danger of the somersaulting tractor – and much else besides. At the rear of the Ferguson's conventional cast wheels, with their double row of steel 'spuds', was a plough. The twin-share plough was not trailed; it was an integral part of the tractor and was firmly attached by a neat three-point linkage, with the depth setting and the raising and lowering movements of the double shares controlled by the driver, who moved a short lever that activated hydraulic rams powered by an engine-driven pump.

This mechanism was a major advance. Not only was it impossible for the Ferguson to rear up but also the shorter length of the plough, and the rigid connection with the positive control it offered, enabled the driver to plough very accurately with crisp furrows as the plough had no wheels that left tracks. The system produced narrower headlands, the tractor/plough being much shorter in length than a conventional trailed plough. Time was saved too; the driver could lift the double shares clear as soon as the end of the furrow was reached, turn to the next stretch, lower the shares, all without stopping the tractor. The Ferguson system was a brilliant innovation, elegantly

Following the deal made between Harry Ferguson and David Brown, the 'Ferguson-Brown' range of tractors was produced between 1936 and 1939. This John Appleyard painting shows the squat foursquare appearance of the tractor very effectively. The one in the background displays the famed 'Ferguson system' of integral ploughs and other implements.

engineered, which converted the farm tractor into a multi-purpose, integrated system, with a wide range of interchangeable implements making it possible for the tractor to become a self-powered plough or any other cultivation machine.

It will possibly come as no surprise that Harry Ferguson, after fruitlessly offering his now patented 'Ferguson System' to various British manufacturers, with such well-known names as the Rover and Morris car companies, Ransome of Ipswich and several others, came to the conclusion that the only way to get the machine into production was to build a demonstration prototype 'Ferguson System' tractor himself. When building the prototype tractor, Harry Ferguson had contracted out some gear-cutting to a specialized family engineering company in Huddersfield. The managing director, David Brown, met Harry Ferguson, possibly to discuss the gear specifications, and became interested in Ferguson's concept.

In 1935, after the 'black tractor' had successfully demonstrated the 'three-link system', David Brown negotiated a contract with Harry Ferguson to manufacture the 'Ferguson-Brown' tractor at their Huddersfield, Yorkshire, works. Production began in 1936. The production tractor looked remarkably similar to the prototype, though finished in light grey rather than sombre black. The engine was different too: the prototype had been powered by an 18hp American Hercules engine. The first 500 tractors off the line had a British Coventry Climax E-series engine, rated at 18–20hp, driving through a three-speed and reverse gearbox.

The Ferguson-Brown tractors with the hydraulic linkage did all that was claimed for them, but sales were disappointing. It is tempting to blame this on an inbuilt Luddite attitude by British farmers, but that is unfair. The Ferguson-Brown tractor was priced at £224, and so was almost double the cost of a Fordson N. It was pointless for the salesmen to say that the Fordson was 'old hat'; it was cheap, a decisive matter in 1937. The Depression had begun in 1929. This had nearly destroyed British agriculture forcing, as it did, thousands of farmers into bankruptcy. There was little money in the 1930s for unconventional, expensive tractors, however brilliant. Indeed, the very innovative nature of the Ferguson-Brown worked against it; the implements required by the system had an additional charge of around £30 each. This sum amounted to fifteen weeks' wages for a ploughman, on top of which, in the market of the time, was a very expensive tractor. Moreover, the farmer's existing stock of trailer implements, which a new, tried and trusted, inexpensive Fordson could utilize, would become redundant.

If the sales had been as good as the advanced Ferguson-Brown system merited, all might have been well. But sales remained too low to be viable; unsold tractors piled up in dealers' showrooms and at the factory. David Brown took the view that a more powerful engine and a four-speed gearbox would improve sales. Harry Ferguson disagreed; he considered that shortcomings in quality control in manufacture were at fault. Unable or unwilling to agree, the relationship between the two men cooled into outright hostility.

Early in 1938, matters came to a head when David Brown set up his own in-house design and development team with the brief to produce a proposal for a revised tractor incorporating the very changes so bitterly opposed by Harry Ferguson. Ferguson retaliated by shipping a Ferguson-Brown tractor, with a selection of integrated

The scene of the famous 'Handshake' meeting between Harry Ferguson (*left*) and Henry Ford at the latter's Dearborn estate. The tractor, carefully posed in the background, is the Ferguson-Brown with the 'Ferguson system' attached which had so impressed Ford.

implements, to the United States to demonstrate the system to no less a person than Henry Ford himself. The meeting had been arranged by Eber Sherman, Ferguson's former business partner in the United States, without, it is alleged, the knowledge or approval of David Brown. The meeting was, for many reasons, to prove to be historic. Ferguson himself demonstrated the tractor and system on Henry Ford's estate at Fair Lane, Dearborn.

The two men, both farmers' sons, got on well. Ford was deeply impressed with the demonstration of the Ferguson system at the conclusion of which a table and two chairs were brought out of the mansion with the Ferguson-Brown tractor carefully posed behind. The famous handshake deal was enacted. This committed Ford and Ferguson to a new tractor, the Fordson 9N, to embody the Ferguson system. Henry Ford had committed his company to an expenditure of millions of dollars and Ferguson had pledged his patents – all on a handshake. It was a magnificent gesture, but no way in which to conduct a multinational business.

A cynic might take the view that the verbal agreement, never to be ratified in writing or even witnessed, took place because Harry Ferguson feared that the small print of the original contract with David Brown might have given cause for litigation against himself and Ford. Predictably, the partnership between David Brown and Ferguson was dissolved acrimoniously in January 1939.

In view of the future ramifications for the British tractor industry, which the sell-out to a ruthless and efficient producer like Ford had undoubtedly compromised, it is fortunate that David Brown, confident in the new design, which was still based on the Ferguson patents, decided to take on Ford and sanctioned full production of the first 'David Brown' VAK 1 tractor. It was launched in July 1939. If that was fast work it was to pale into insignificance; just eight months after the first meeting with Harry Ferguson at Dearborn, the Ford 9N was in full production.

In June 1939, pre-production 9N tractors had been demonstrated and filmed on Ford's Dearborn estate to promote the Ferguson system, the most impressive being the ploughing of an area the size of an average back garden. This plot of just 20 x 27ft (6 x 8m) was ploughed without any wheel marks remaining after the tractor had left. The tractor had the Ford logo (Fordson seems to have been dropped in the United States) on the radiator with a second plate just below reading: 'Made in USA' above 'Ferguson System'.

The Ford-Ferguson, as it was to become known, was a very good tractor indeed. It utilized the complete Ferguson three-link hydraulic linkage with both Ford and Ferguson engineers collaborating on each and every detail. A new Ford side-valve petrol engine was developed, offering 23.5 rated hp at 2,000rpm, a three-speed gearbox and, of course, a full range of hydraulically operated implements, all refined and mass-produced at Dearborn. During the eight-year production life, halted for a time by wartime restrictions in 1942, more than 300,000 9Ns were to be produced. It was very popular with farmers, partly due to the Ford policy of low price.

In the United States, the Ford 9N proved to be very successful, with over 40,000 units sold in the first year. A small number had been exported into Britain, and were well received. Harry Ferguson, therefore, had reason to expect that the English branch of Ford at Dagenham in Essex would be interested in producing a British version of the Ford 9N with the Ferguson system. But this was not so. By now war production was stretching the industrial resources of the country. Raw materials, especially high-grade steels, were at a premium with munitions and aircraft factories having first claim on them.

The British government did not make the same mistakes that had so nearly starved the country in 1917. German U-boats, up to 1942, again sank an appallingly high number of ships, but this time agricultural production was on a war footing; skilled farm workers of military age were exempted from call-up as their contribution to the war effort was higher in farming than they could offer as servicemen. The Women's Land Army was reintroduced, and even POWs were working on the land. The tractors they drove, however, were the trusty Fordson Ns.

Despite meetings between Harry Ferguson and the Dagenham management with, very possibly, strong hints from Dearborn, the British Ford tractor which, in the postwar period, would replace the Fordson N (of which over 750,000 had been produced) was not to be the Dearborn Ford-Ferguson 9N, but an improved and cosmetic redesign of the old Fordson N. It was designated the Fordson E27N Major. The basic design dated from 1916, and it did not incorporate any of the Ferguson system patents. The petrol engine of the Fordson E27N Major was still a side-valve unit but developed 30hp. Pneumatic tyres were offered with other improvements but, once again, the strongest selling point was price; the basic version was a highly competitive £237.

In 1948 Ford offered the option of a Perkins diesel engine. The Fordson E27N Major was popular and was to remain in production until 1952. By that time, the simple, rugged, very reliable and inexpensive Fordson had been in production for an unbroken thirty-six years – a total of over a million tractors. If the inability on the part of Harry Ferguson to persuade the Dagenham board to build a British 9N was a setback, much worse was to follow.

The famous handshake partnership, the precise details of which remain obscure, held. Indeed, Henry Ford authorized a new range of Ford-Ferguson tractors, to be designated Ford 8N, to meet the expected postwar demand. But in 1945 Henry Ford retired and Henry Ford II took over control of the company. One of the problems he had to address was that the sales of the Ford-Ferguson 9N had diminished to the point of unprofitability. A new design was urgently needed. Attempts were made in 1946 to negotiate a new agreement with Harry Ferguson without success. Harry Ferguson found Ford's proposed marketing arrangements unacceptable, and Ferguson demanded the sole rights to these as the core of a new agreement. That being the case, Henry Ford II, faced with the responsibility of investing tens of millions of dollars mass-producing a new tractor, took the position that he was not beholden to honour the original verbal agreement made with Harry Ferguson in 1938, which was the legal position (as Sam Goldwyn once memorably remarked: '. . . a verbal agreement is not worth the paper it is written on'). In November 1946, without any involvement on the part of Harry Ferguson, a new company, Dearborn Motors, was set up as a sales company to market a successor to the ageing Ford 9N, confusingly designated 8N.

Henry Ford II, as Harry Ferguson saw it, had reneged on the 1938 deal by producing and marketing the 8N tractor under Ford's total control, thus severing all connection with Harry Ferguson, even though his patented Ferguson system remained a feature of the new 8N tractor. A protracted lawsuit followed in which Harry Ferguson claimed compensation amounting to $344,000,000 – a gigantic sum. The suit, reminiscent of Jarndyce v. Jarndyce in *Bleak House*, would drag on for four years.

Attempts were made by Ford to settle out of court, but to no avail. Eventually a settlement was arrived at with Ferguson winning the case, but only getting a paltry $9.25

The tractor of the 21st century? The JCB company decided to reinvent the agricultural tractor and the 'Fastrac', introduced in 1991, certainly has many innovative features: full suspension which is self-levelling, powerful engine options (115–170hp), engines which offer a road-speed of up to 45mph (75kph) and a full range of computerized electronics. Uniquely, the comfortable cab can accommodate a passenger.

An action shot of one of the 300,000 'Grey Fergie' Ferguson TE tractors. The integral plough is well illustrated, as is the case with which the driver can adjust or lift the shears without leaving his seat or having to stop the tractor.

million in agreed damages. Even this figure was reduced by legal costs to $6 million.

Harry Ferguson was determined to produce a British tractor to his design incorporating his system. Clearly, Ford in the UK was not acceptable so a new manufacturer was sought. He was able to conclude a deal – in writing, one suspects – with Sir John Black, the managing director of the Standard Motor Company of Coventry.

The result was the famous TE20 'Grey Ferguson'. This small, very well-designed and well-engineered all-British tractor was to become a bestseller, both at home and abroad. It incorporated the Ferguson system sold under the slogan: 'It's what the implement does that sells the tractor.'

The first tractors designated Model TE20 (the letters standing for 'Tractor England') left the Standard factory in Banner Lane, Coventry, in 1947. The 28hp tractor was offered with the possible use of no fewer than forty 'Ferguson Implements'. It had a four-speed and reverse gearbox, and the very small turning circle of 17ft 6in (5.25m). The weight of the petrol version was only 1 ton 2cwt (1.12 tonnes), and the price was a competitive £325. By 1952, a 26hp diesel engine was an available option.

Over 300,000 TE 'Grey Fergy' tractors were made at Coventry between 1947 and 1953 when Ferguson, in a complex series of deals in which Harry Ferguson was paid $16 million in Massey-Harris shares, merged with the Canadian-based agricultural machinery company, Massey-Harris. Production was changed to Massey-Harris at Detroit, the American-built tractors being then designated Massey-Ferguson, Model TO20. In 1957, following the perhaps inevitable deterioration between the Massey-Ferguson board and Harry Ferguson, he offered to resign. The offer was accepted; Ferguson sold out his share interest in the company.

In 1958, Harry Ferguson was considering yet another tractor development but the motor industry in Britain was cool and the proposition came to nothing. So ended Harry Ferguson's lifelong involvement with tractors. His 'Grey Ferguson' is without doubt a tractor classic, to be compared to the trend-setting Fordson F. Seeking a fresh challenge, he formed a new Coventry-based consultative company, Harry Ferguson Research Ltd, and turned his attention and considerable talents to his early interest, Grand Prix cars. The Ferguson Project 99 was a unique four-wheel-drive Grand Prix car which, driven by the young Stirling Moss in 1961, showed considerable promise. But by then, in October 1960, Harry Ferguson, who had for some time been in failing health, had died at the early age of 58.

His legacy can be seen to this day on the farms of two continents. The 'Grey Fergusons' are still working in considerable numbers and the system they use is, in a developed form, now universal. The effect of the tractor on the mechanization of agriculture in the Western world is immense, as the writer and broadcaster Quentin Seddon has pointed out. He notes that in 1929 there were half a million horses at work on farms in Britain but, twenty-five years later, tractors outnumbered horses. By the 1970s tractors outnumbered farm workers. Today there are some half a million tractors on the farms of this country – the equivalent of 25 million horses.

The contemporary tractor is far more powerful and versatile than any of the preceding classics. Tractors now have comfortable cabs with seats designed with expert medical advice, air-conditioning, heaters, and VHF radios for communication and broadcast listening. They have turbo-diesel engines up to 200hp with electronic control, automatic gearboxes, satellite-positioning units, four-wheel drive, disc brakes, electronic control of the hydraulics, power take-off for implements and many other refinements. But the basic concept of their shape and integrated multi-systems is by Ferguson out of Fordson.

What are the candidates for the classics of the future? This is difficult to foresee. The essential point for the creation of a true classic is that it has set a trend which others merely follow and develop. Daniel Albone's Ivel has a place because it was the first practical tractor; it showed that the internal combustion engine had a role in the mechanization of agriculture. The Fordson N is a true classic because it defined the shape and basic function of the farm tractor, which has remained valid for over eighty years. It had the 1:3 ratio of front to rear wheels and was an affordable machine that could outperform the functions of the horse teams it replaced. The Ferguson system, as applied to various tractors – in particular his own TE20 series – set a trend which, in developed form, remains the norm. There will, no doubt, be many other claims to classic status in the years to come.

One that might be included in a short list is the JCB 'Project 130', the Fastrac. This state-of-the-art tractor has a shape that owes little to the past: it has all the features required in the way of high-tech mechanics and electronics. And, vital for a potential classic, they look different from all that have gone before. Developed in the 1990s at a cost of £12 million, the company has high hopes of the radical Fastrac design. JCB's Mike Butler says: 'When we started we thought we had found a niche, but this is not a niche market at all. It could turn out to be the most important development in agricultural machinery since Harry Ferguson put a three-point linkage on the back of his tractor.' A classic in waiting? Only time can tell.

ROADS AS WE KNOW THEM today were the invention of the Romans during their 400-year occupation of Britain. Before the Roman invasion, in what is known as the Iron Age, settlements were connected by tracks, maintained simply by use and the constant passage of people. These would have been drovers and herdsmen with their animals, horses, oxen and wheeled wagons. When excavated, most of the surviving tracks discovered are about 5ft (1.5m) wide, which was the average width of the wagons. In later centuries the major tracks were widened to permit two wagons to pass. Wide or narrow, these ancient tracks, whenever possible, followed the contour lines of the higher, well-drained and therefore firm ground of the countryside. They avoided forests and dense woodlands which were, for many reasons, considered to be dangerous to the lone traveller. Swamps and boggy ground were also avoided. Eventually, following centuries of continuous usage, there emerged a clearly defined, countrywide network of interconnecting tracks that were in no sense planned or constructed; they just happened, like the tracks formed by wild animals.

These pre-Roman tracks had one very serious limitation: weather. There was no top dressing of any kind; the surface was simply the beaten earth, be it chalk or clay. In winter the tracks could be a sea of mud, and a bad winter storm could wash a track away, while in high summer the traveller would be contending with clouds of dust. If a track was unused for any length of time, because of a landslip or fallen trees, it could well be lost for ever as nature took over. Primitive though they were, the tracks served well enough the needs of the time. People travelled little. Travelling could be very dangerous: wild animals still roamed the countryside, as did bands of outcasts who would kill a man for the clothes he wore; there were no inns; and local communities were enclosed, suspicious of all strangers and reluctant to allow them to pass through their area, still less to spend a night within the community. One way and another, there was little incentive to journey far. Settlements were very largely self-supporting and withdrawn, though in later centuries the growth of towns and ports encouraged merchants to travel in pursuit of trade.

Before the Roman came to Rye
or out to Severn strode,
The rolling English drunkard made
the rolling English road.
(G. K. Chesterton)

When the Romans invaded Britain in AD 43 with their four legions (II, IX, XIV and XX) they had different imperatives. They were occupying and subjugating a hostile, wild land and needed, above all, good military communications. As the legions advanced,

Pilgrims Way in Kent remains one of the few ancient tracks which can be walked at length. When this photograph was taken in 1951, fifty-three miles had just been made passable. Had they walked it today, no doubt the Knight, the Miller, the Monk, and the wife of Bath *et-al* would have had different tales to tell.

military engineers built roads behind the fighting soldiers as soon as the land was conquered. Roads were needed for supplies, reinforcements and in the remote event of the soldiers having to withdraw back to their fortified bases in order to regroup. Once the initial occupation had succeeded, the newly constructed roads offered the facility to move troops quickly from one place to another to counter possible insurrection, and to be able to supply and maintain distant garrisons. The Romans clearly did not consider the existing British tracks adequate, even those few which happened to be leading in the desired direction. The ancient Britons made tracks; the Romans built roads.

The Roman military engineers and surveyors responsible for the planning and construction of their road system, although lacking in any form of mechanized aid, nevertheless had one very great advantage over their modern counterparts: they did not have to contend with public enquiries, planning laws, sites of special scientific interest, landowners with cast-iron manorial rights and people living in trees or burrowing beneath the proposed roadway in protest. Any ancient Briton who wished seriously to impede or dispute the right of way of Rome would find himself incorporated into the *pavimentum* before he could demand '*Quo vadis*?' The lack of political restrictions enabled Roman road engineers to construct their straight roads through ancient barrows, farms, settlements and fields of standing crops, only deviating when the nature of the topography made it essential – perhaps skirt a river, swamp or steep escarpment. It is said that the Romans kept all curves as gentle as possible because, oddly enough, Roman wagons did not have steerable front wheels and the fixed four-wheel chassis had to be slewed around corners. This made it difficult, if not impossible, for the wagons to take the sharp, tight bends that are a feature of a zigzagging road, which winds across the face of a steep hillside to ease the load for draught animals when climbing up. (The pivoted axle did not arrive for another thousand years – early in the sixteenth century.)

Roman roads in Britain had a width of 14–16ft (4.2–4.8m), wide enough for two columns of soldiers, six abreast, to pass. The paved tracks for wheeled vehicles have been measured by archaeologists at 4ft 8in (140cm) (significantly, the international standard railway gauge, invented by Stevenson, is 4ft 8½in (141.3cm)). Once the initial fighting had ceased and the country had been settled, more or less, the military road network was enlarged to the finalized extensive system, the major elements of which, though largely buried beneath modern concrete, still exist. It consists of Watling Street, Fosse Way, Ermine Street, Stane Street and many others. All the major Roman roads were planned meticulously before any work was commissioned. Once the route had been approved by the provincial governor, military surveyors or *gromatici*, decided on the precise alignment of the new road and erected survey posts to define the route for the builders. All measurement was made in the Roman mile: 1,000 five foot paces of a running man; this is near enough to 1,620yd (1,480.6m). It seems that the Romans did not use forced labour to construct their roads in Britain, though paid local labour might have been used. The road was built in sections by troops from local garrisons.

The sequence of building a Roman road was much as that used today. First the route would be cleared of trees and undergrowth. A certain amount of levelling would be undertaken with excavation cutting through rising ground or the construction of an *agger* (embankment) if the ground level was falling below the planned level of the roadway. All digging, clearing of vegetation and tree felling would, of course, have had to be done manually. For the military engineers it was a well-known routine; by the time the Romans built roads in Britain they had accumulated many years of road-building experience elsewhere, and so Romano-British roads were cambered for drainage and properly engineered. First came the foundation, the *pavimentum*, then followed a second layer of stones cemented by mortar, the *statumentum*. Next came the *ruderatio*, the third layer, consisting of smaller stones cemented together with mortar; on top of that was the *nucleus*, which was a layer of crushed stone, chalk, sand or lime depending on the availability of local supplies. Finally the *dorsum* was laid down; this was the top dressing, usually a paved surface. Kerbstones might be added in certain places, in addition to drainage ditches running either side of the finished road. The Romans provided no maps or signposts of any description; the traveller just had to know or ask the way.

The final extensive network of major and minor Roman roads served Britain for the 400 years of occupation. Although the Roman official and military traffic had total priority of usage, the local population was permitted to use the roads but woe betide anyone who impeded official traffic. It is said that the drainage ditches were unintentionally kept clear by the passage of British travellers forced off the road to make way for Roman traffic. The roads were free; no tolls were raised but wealthy landowners whose land was crossed by the road were often required to make a contribution to its upkeep.

For 400 years the roads were maintained by the Romans; when they left Britain in the fifth century Roman roads, like the Roman Empire, fell into decline. Wooden bridges soon collapsed; local people looted stone from the roads for building, and the weather and vegetation soon took their inevitable toll. Even before the Romans had departed, shortages of troops forced them deliberately to destroy roads and bridges in East Anglia to impede the invading Saxons; they were never to be rebuilt. For over a thousand years after the departure of the Romans there was no significant road building in Britain. The disintegrating original Roman roads and the ancient Iron Age tracks remained all that was available.

Some sections of minor Roman roads exist, still in surprisingly good original condition. There is Akeman Street near Cirencester, Gloucestershire, Ackling Dyke near Sixpenny Handley, Dorset, stretches of Stane Street in and near Eartham Woods, Sussex, and many others. From the top of Birdlip Hill the modern A417 from Gloucester to Cirencester follows exactly the alignment of the original, arrow-straight Roman road along which soldiers of the II Legion once marched. This is but a fragment of the communications of a great, vanished empire, along which marched the legions fulfilling the required ten miles (16km) per day and the couriers of the *cursus publicus*, the imperial post, carrying reports from distant provinces, orders, admonishments, promotions and news from Rome. From Hadrian's Wall south through Britain, through Gaul and Iberia to North Africa, along the Mediterranean coast to Lepcis Magna in Tripolitania, through Cyrenaica to Tripoli and Alexandria, east to Petra, Antioch, north to Byzantium, then following the course of the Danube and the Rhine, all roads led to Rome. These were the motorways of antiquity.

By the sixteenth century the state of major roads in Britain was appalling. Trade was flourishing and merchants required heavily loaded carriers' wagons, some drawn by as many as ten horses. These wagons were fitted with wheels with very wide rims, the idea being that such wheels would level the earthen road surface. They did nothing of the kind – far from levelling the roads, they sank into the soft surface, producing deep ruts which filled with liquid mud and water in winter and were as hard as concrete in summer. The unsprung wooden carts were shaken to pieces. The government did what most subsequent governments did: instead of improving the roads they passed punishing Road Acts to limit the use of the existing deplorable highways. From 1751 no fewer than sixteen acts were passed which defined the size, loading, weight of wagons and the nature of the wheels they used. There were even suggestions in government circles that major roads should legally only be open to pedestrians and horsemen; wheeled traffic should be banned altogether. Fortunately, there were sufficient members of parliament with trading interests to discreetly squash the idea. The Road Acts did succeed in limiting the size and weight of wagons, and there was some work done to consolidate existing roads using funds raised from local turnpike tolls. The roads were marginally improved, with fewer potholes and smoother surfaces. Suspension of the body of coaches, firstly by leather straps which offered a rudimentary form of springing and then, in the seventeenth century, by the adoption of the metal 'C-spring' at least made possible the era of the stagecoach and, for the wealthy, fairly comfortable travel in lightly built carriages. In 1784 the first Royal Mail coaching service was inaugurated between London and Bath, replacing horseback riders. Even so, the roads were still far from good, and the speed of long-distance travel was no faster than it had been in Roman times. As more people needed to travel, the increase in traffic, particularly that of the influential 'carriage folk', led to demands for a major new road-building programme. 'Comes the hour, comes the man', or in this case two men: Thomas Telford and John Loudon McAdam.

Thomas Telford (1757–1834) was an engineer who built the suspension bridge across the Menai Straits between North Wales and Anglesey; he also designed 900 miles (1,440km) of new roads. Telford believed, as had the Romans, that a good solid

foundation was the key to a long-lasting road; the one he built to Holyhead was considered at the time one of the most outstanding roads of the coaching era. McAdam (1756–1836) was a Scottish engineer who specialized in the design of the good, cambered, weatherproof surface that was to bear his name and which was achieved by compacting a 12-in (30-cm) layer of graded small stones on the subsoil and consolidating them by pressure. Traffic and usage provided the main source of further consolidation. McAdam's structure withstood considerable wear without rutting. This 'macadam' formed the main surface of roads from the coaching days through to the Industrial Revolution, when some pressure was taken away from the main highways by the extensive railway system which, by the middle of the nineteenth century, was as long in mileage as the major roads. Despite the developments by Telford and McAdam in construction, there were few really new 'green-field' roads (i.e. ones where no road previously existed); many were the result of simple resurfacing and widening of either existing Roman roads or the ancient Iron Age tracks – the 'Rolling English Road' of G.K. Chesterton:

Before the Roman came to Rye or out to Severn strode,
The rolling English drunkard made the rolling English road.
A reeling road, a rolling road, that rambles round the shire,
And after him the parson ran, the sexton and the squire;
A merry road, a mazy road, and such as we did tread
The night we went to Birmingham by way of Beachy Head.

Greenwich to Charing Cross, 18th-century style. By that time, the growing use of stage coaches – though few would have been as well patronised as this one – had revealed to a growing number of travellers the deplorable state of the roads in England. The subsequent improvements by Telford and McAdam were to be forced on the government by public outrage.

Throughout the nineteenth century English roads, rolling or otherwise, changed little; macadam formed the surface of the main roads. In the towns the dramatic increase of traffic caused by cabs, wagons and horse buses was countered in part by widening existing streets, though the surfaces were either macadam, cobblestones, stone sets or wooden blocks covered by tar (which soon wore away because of the almost continuous action of iron-tyred wheels, with the exposed wood, when wet, becoming very slippery indeed). The arrival of the urban tramcar with its tracks was an additional hazard. There was no plant as such to aid the construction and repair of the urban streets, the only marginal development being a coal-fired tar boiler drawn by a horse to the site. Stone sets were laid one by one by hand on a sand base and rammed with heavy wooden rammers until level. Tar was then poured into the intersections of the stones, rather like the mortar of a brick wall. Wooden blocks would be laid in the same way, finally being covered by molten tar. The surface of macadam roads and streets was levelled by the use of hand- or horse-drawn rollers, though the weight of such rollers was limited as otherwise they would become uncontrollable on hills. In 1865 the first mechanical plant arrived, which was to transform road works: the steamroller. The steamroller was to prove long lasting; from its introduction in 1865 it was to survive to work into the 1960s and the era of the motorways – truly 'classic' plant.

Strangely, in view of the fact that the steam railway locomotive and the traction-engine were British inventions, the steamroller was invented in France in 1860. But French rollers were, at 30 tons, too heavy. Thomas Aveling, a farmer and amateur engineer from Kent, had hired a portable steam engine for threshing. Aveling decided that the horse-drawn portable engine might be made to propel itself. He sketched out his ideas and took his proposal to the makers of the portable, Clayton & Shuttleworth, who surprisingly did not dismiss the design out of hand but built a self-propelled version of their engine to Aveling's requirements. Its performance must have encouraged him

The first Aveling & Porter steamroller of 1865 was, at 30 tons, too heavy. This was in part due to the 500 gallons of ballast water in the large tank behind the intrepid driver, who can be seen using a ship's wheel to steer. 'Could do better' might have been the verdict. Aveling & Porter did: their revised 1873 machine was definitive, with the roller in front and the weight down to around 10 tons. It was to be followed by 8,600 more.

'By Appointment.' Two Aveling & Porter steamrollers working to level the parade in front of Buckingham Palace on an unknown date during the 1930s. Just why it was considered necessary to use two for this relatively modest area is unclear. Were the contractors worried that one might break down? Or was it a question of getting the job completed before the Changing of the Guard?

for, in 1862, with a partner, Richard Porter, he founded the firm of Aveling & Porter at Rochester, Kent. Three years later they had produced their first steamroller. The trade mark of the company was the rampant horse of Kent above the scrolled motto 'Invicta', which was to grace the front of the first and subsequent 8,600 Invicta steamrollers that were eventually to be produced by Aveling & Porter.

The first steamrollers that the company built, were, like the French machine, very heavy at around 25–30 tons, in part because of a large 500-gal (2,250-l) water ballast tank over the main roller which was at the rear of the machine; this heavy roller was steered by a ship's wheel. The layout, though consisting of three rollers, was the reverse of subsequent practice. By 1873 Aveling & Porter's steamrollers had been revised and emerged in the classical three-wheel form with the cast bracket bolted to the boiler casing to support the front roller. This was actually divided into two sections to give a differential effect when turning. The weight had dropped to the more modest and practical 8 tons. Aveling & Porter developed their single-cylinder 'simple' steam engine into the far more efficient twin-cylinder 'compound' type, thus following the latest practice of railway locomotives. In a compound engine the steam is used twice; once in a high-pressure cylinder that exhausts into a second, low-pressure cylinder which in turn exhausts into a blast pipe at the base of the chimney to draw the fire. The Aveling & Porter steamrollers were excellent performers and in great demand, principally by county councils who were responsible for the upkeep of roads in their county. (This applies to this day. On most country roads it is noticeable that where the road crosses a county border, there is often a marked change of surface reflecting that county's civil engineer's ideas – or his budget.)

By 1900 the success of the Aveling steamrollers was such that most traction-engine makers decided to offer a steamroller. Well-known names such as Fowler, Clayton & Shuttleworth and Burrell & Marshall all followed the, by now, traditional classic form

The pneumatic tyres reveal this Sentinel steam tanker to be a late production c.1938. Beneath the tar tank can be seen the pumps and associated piping for the steam-heated coils which keep the tar in the tank molten. W. J. Glossop of London were major contractors with a large fleet of Sentinels.

which was to remain, virtually unaltered, until the steamroller was finally displaced by the diesel roadroller in the 1960s.

At the end of the nineteenth century, the techniques of road building and repairing were changed with the adoption of the steamroller. Many carried a scarifier at the rear of the engine. This consisted of three or four strong steel picks or tines which, when lowered to the road surface, would dig in to 'scarify' or tear up the old, damaged surface. The rubble would be cleared away by men with shovels who then manually levelled the surface. If the road under repair was secondary, a water-bound technique would suffice. Broken or small stones would form the base, and this would be compacted by the steamroller trundling back and forth until the area to be treated was level. Sifted earth was then added which was watered to a slurry, then rolled level and left to dry. When dry, a hard level macadam-like surface was produced. This Victorian version of the macadam road served until the arrival of the motor car and, more specifically, the heavy lorry and omnibus. These relatively high-speed vehicles, in summer, threw up dense clouds of dust. In towns this was treated by water carts spraying the road surface to lay the dust – hardly a practical proposition on country roads.

The water-based macadam road was, in the years before the First World War, improved by the use of bitumen as a binder. Bitumen, a black hydrocarbon compound, was, when mixed with limestone and sand, found to be an ideal medium to bind aggregate in order to form a resilient waterproof road surface. Bitumen, once known as 'Jews' Pitch', is a generic term dating from the fifteenth century; the substance occurs naturally from the distillation, by evaporation, of certain crude oils and lignite. It was known to the ancient Egyptians, who used it to waterproof and protect the outer covering of embalmed mummies. It was the discovery, in Trinidad in the West Indies, of extensive lakes of a form of bitumen known as asphalt which made it available to be used for road surfacing in sufficient quantities. Asphalt that is mixed, like bitumen, with sand and limestone is solid at normal temperatures but, with heating, readily becomes semi-liquid and can be poured and spread, quickly solidifying to a smooth, hard surface.

The other vital aid to road works is tar. In appearance tar is a similar sticky jet-black substance to bitumen, but it is man-made. Tar is obtained by heating coal to around 1,832°F (1,000°C) in an oxygen-free sealed container. Liquid coal tar is the result. It is rich in organic chemicals, including benzene, which are produced by additional distillation. The tar produced from coal is used just as it comes out of the heating

Roadworks: a lightweight steam roller at work on a secondary road during the 1930s. The men with the brushes are spreading stone chippings evenly over molten tar as the roller consolidates the newly laid surface. The divided front roller, which offered a degree of differential to prevent scuffing when the steamroller was turning, is well illustrated.

A section of a suburban road in Lincolnshire being relaid with 'Tarmac' in 1938 when some major contractors, like Val De Travers, had turned from steam to diesel roadrollers. A pile of fresh tarmac has been tipped by a lorry and the men are loading it with forks into wheelbarrows before spreading it on the road surface. The section which has been rolled can clearly be defined.

process when it then cools to a rich-black, resilient solid. With re-heating this becomes once more a free-flowing liquid. This hot liquid, poured over a prepared macadam surface, could be dressed by spreading stone chippings over it and compressing the whole with a steamroller. The tar set, leaving the stones embedded in the tar to form a reasonably hardwearing, waterproof, dust-free surface, the technique being know as tarmacadam. One drawback that was discovered was the low melting point of the underlying tar, which could become semi-liquid when heated by prolonged, hot sunlight. However, that condition was rare enough in Britain not to form a serious drawback. (The technique is still in use on secondary roads today. Wise local drivers avoid such newly treated roads for a week or two, until the surplus stone chippings have been swept up or embedded in the tyres and paintwork of the unwary.) In Victorian times, the stone was produced manually by old men and even children breaking stones into chippings at the roadside with hammers. After the First World War steam-powered crushers produced chippings commercially for roadworks.

The First World War had witnessed a certain limited amount of new secondary road building to serve the needs of the army. After the war, the 1920s and 1930s were a time of considerable revision of Britain's road network. Principally, this involved the widening of urban roads and streets to try to cope with the very large increase in motor traffic of all descriptions on roads that had evolved for the use of horse-drawn vehicles. Major trunk routes were also widened for the same reason, and extensively relaid with water- and dust-proof surfaces based on tar and asphalt. The sight of roadworks became a familiar one; usually there would be a red sign warning of 'Road Works Ahead', nearer there was possibly an additional warning of 'Steamroller'. The actual site of the relaying of that road section was often surrounded with wooden scaffold poles draped with red oil lamps lit at night. There would be many men working, some tending a coal-fired tar boiler. There might be a boiler melting asphalt and mixing it with sand and limestone; the mixer boiler had an internal paddle to perform the mixing, which was slowly turned by a wheezing single-cylinder diesel engine. The tar was usually applied by men using what looked like large watering cans. It was then evenly spread by other men with stiff brushes, and a top dressing of stone chippings was skilfully applied by a gang with shovels using sweeping movements to ensure an even layer of stones spread on the surface of the molten tar.

When the foreman was satisfied that the section had been satisfactorily prepared, the steamroller, simmering nearby, chugged into rumbling action to compact and level the road surface. If the section of road under repair was large then a second classic plant, a magnificent steam vehicle such as a Sentinal or Fowler tar tanker, would be present. The tar tanker had a large tank on the chassis with steam fed through the coils (which became hot) from the boiler to keep the tar at 356°F (180°C) and fluid. At the rear of the tanker there was a spray bar with several spray heads through which hot molten tar was pumped on to the prepared road bed. The flow of tar was at a constant rate, the depth being regulated by the forward speed of the tanker: if the speed was slow, deep layers of tar resulted; if the speed was faster, the layer would be thinner. The drivers of these, usually contracted, tankers were very skilled and could judge to fine limits just how much tar should be applied. When a tar tanker was in use the chippings had to be available in large quantities. If all went well the steamroller compacted the newly laid surface so that it was set into a smooth, level, waterproof, dust-free road.

The men who worked with the steamrollers and the tar tankers led a hard life. One of them, Alan Pronger, drove an S-class Marshall steamroller, which was owned by West Sussex County Council, for twenty-years. The roller was named 'Joan' after Alan's wife. When working away from base he would eat and sleep in the 'living van' which the roller towed, together with a water cart, from site to site. At the weekend Alan and other steamroller drivers would go home, but:

> At the start of the week, you had to be early on the job on Monday morning. I used to return on the Sunday night because the roller was stone cold and it took a long time to raise steam. I would get up at half-past five to light up the roller, cook and eat my breakfast, then go out to the roller and oil round . . . You had steam up and were on the job by half-past seven when the gang arrived. That routine went on week in, week out.

Each night Alan Pronger and hundreds of other steamroller drivers up and down the country would, if working too far to cycle home, spend their nights in the living vans. These were wooden and contained bunk beds, a coal-fired kitchen range for cooking, a table, oil lighting and buckets for water. They did not offer much else in the way of domestic comforts. In the winter it was far from an agreeable life. Alan Pronger recalls that:

Alan Pronger, who drove steamrollers for many years, has recounted how he had to live an itinerant lifestyle working far from home. A typical 'living van' – now preserved – in which many road workers had to eat and sleep, is being towed behind the steam roller.

> . . . when you woke up in the morning you found your bedding frozen to the side of the van and thick ice on the drinking water in its bucket. And the lid frozen to the kettle. It was pitch dark too . . . In summer it was lovely of an evening sitting in the van probably peeling potatoes for the evening meal and giving the van a sweep out or perhaps polishing the kitchen range.

Alan Pronger also remembers that, before the Second World War, when repairing a secondary road the gang would see perhaps two cars pass the site in half an hour.

In the 1930s roads were surfaced or 'metalled' with crushed slag from iron- and steel-works. The slag had been pre-coated with tar known under the trade name of 'Tarmac'; this arrived on a lorry, ready mixed from the depots, and was tipped into a hot, malleable heap beside the road. Men with forks that had closely spaced tines dug from the pile and spread the tarry aggregate into place. The steamroller then levelled it, men standing by with shovels full of tarmac to fill depressions as the roller worked. To prevent the sticky, tarred stones clinging to the rollers, most rollers had a water spray that kept the rollers wet and discouraged the tar from sticking. There were, in addition, scrapers fitted with minimum clearance on the rollers to remove any stones that might be sticking to them.

The history of 'Tarmac' is an interesting one. In 1901 a Mr E. Purnell Hooley, then Nottinghamshire's county surveyor, happened to be passing by a local ironworks when he noticed that, some time previously, a barrel of tar had apparently fallen from a horse-drawn dray, had burst open, and had spilled the liquid tar into a large puddle across the road. This accident seemed to have been reported to the ironworks for, to prevent further mess, the tar had been coated with waste slag which had been absorbed by the tar. The observant Mr Purnell Hooley noted what no one else had: the tar and slag puddle had solidified to form a hard, dust-free, unrutted surface to the road.

The surveyor seized on the idea of mixing waste (and therefore cheap) slag with molten tar. After trials to obtain the correct ratio, he patented the mixture under the trade name 'Tarmac', presumably a contraction of the existing, though unregistered, Victorian road dressing 'Tarmacadam'. Although the idea was perfectly sound, Mr Purnell Hooley should have stuck to his theodolite; he lacked the business skills to raise sufficient capital successfully to market the invention. There were others who had both. One was the owner of extensive Midland steelworks, the MP for Wolverhampton, Sir Alfred Hickman, who in 1905 bought the Tarmac name and relaunched the company as Tarmac Ltd, based in Wolverhampton close to virtually unlimited supplies of iron slag. The First World War provided the opportunity for business as the government bought shiploads of Tarmac to lay military supply roads over the mud of Flanders for the British Army. To maintain supplies, Tarmac Ltd were allegedly even allowed the use of German prisoners of war to work on the production of Tarmac, replacing many of the Tarmac company's own workers who were by then marching along the 'Tarmac-metalled' military roads in France into battle. After the war, to ensure continuing supplies of slag, Tarmac Ltd bought up entire iron slag tips and roadstone quarries and set up depots and works around the country to supply their product for local roads. The Tarmac company diversified into civil engineering and flourishes to this day – all because a drayman stowed a barrel of tar carelessly in 1901.

The other item of plant available for road building and repair after the First World War was the very unpopular pneumatic drill with its noisy compressor, usually built by a company named Broomwade. The drills produced an ear-splitting, machine-gun-like, staccato noise that penetrated office and home alike. Noisy they certainly were, but they did ease the labourers' job in digging up old, damaged road surfaces. (They remain, in a slightly silenced form, still with us.)

Although the pneumatic drill helped, steamrollers and their crews remained the prime plant available for roadworks before the Second World War. Practically every county council in Britain owned steamrollers, some with fleets of over a hundred. Contractors also offered their services, ranging from one man with his single roller to quite large companies, many offering traction-engines too. It has to be said that the steamroller, though now an object of loving – and very expensive – restoration and preservation had, from 1927, a serious competitor: the diesel roller.

The motor roller goes back a long way. The *Engineer* magazine of March 1902 reported on a French petrol-engined roller while the English company of Barford & Perkins produced a motor roller of 3 tons, probably for the use of sports grounds, around the same time. There were others, all with paraffin or petrol engines, which do not seem significantly to have threatened the supremacy of the steamroller until the first diesel roller, from Barford & Perkins, appeared in 1927. Although the diesel roller was more efficient than the steamroller, it did not readily displace it, one reason being that steamrollers were built to a very high standard, so over-engineered that they simply went on for ever. Also, there were so many of them; a county council with a hundred or so would not lightly contemplate replacing perfectly usable steamrollers when good Welsh steam coal was cheaply available between the wars. One contractor, E. J. Sandercock, has written that a new steamroller could be expected to have a minimum ten-year working life before any major repair was required; diesel rollers, on the other hand, needed the constant maintenance of injectors and fuel pumps together with an engine that lacked the rugged simplicity of the steam engine. Eventually, the rising cost of steam coal and the need to have supplies of soft water for the boiler tipped the economics in favour of the diesel. However, steamrollers hung on for many years and there was always work for them. By 1938 there were 125,000 miles (200,000km) of roads in Britain; most had been more or less relaid since the end of the First World War in 1918 and most of those miles of road had been levelled by steamrollers – a very considerable achievement.

Among all these miles of prewar roads were some genuinely new ones. These were the famed 'bypass' roads built to relieve the growing congestion in town centres; a laudable aim nullified by the lack of planning of the time, with the result that the bypass roads became heavily built up by ribbon development, which was the unregulated building of houses and shops along much of the new bypasses. These were roads for the motor age; they had dual carriageways and were illuminated at night by the new sodium vapour lights. There were no speed restrictions and the roadway was surfaced with a material new to British road builders: concrete. Ironically this, too, had been known to the Romans although they did not use it for their road building.

Concrete is a mixture of sand, stone aggregate and cement or lime, which permanently sets by chemical reaction to a very hard, waterproof and durable artificial stone; it is inexpensive to make and lay. The wet mixture is poured on to the prepared road bed and confined by wooden or metal shuttering; before it sets, a long beam is laid in contact with the still wet surface. A small petrol engine mounted on the beam with an eccentric coupling causes the beam to vibrate, thus settling the concrete into a level, characteristic ripple finish, which offers good adhesion to tyres in wet weather.

Bypass building was only on a restricted scale in Britain. From the middle of the 1930s in Germany a very extensive programme of 'green-field' road building was put into operation; it was a nationwide network of concrete dual-carriageway super-highways that the Germans called *Autobahnen*. The Nazi government claimed that the new roads had been built solely to relieve unemployment, which was very high in Germany at that time. The road-building programme did relieve unemployment but that was a fringe benefit. The autobahn had been built for quite another reason; like the Roman roads centuries before, the new German super-highways were routed and built for the express purpose of moving troops and military supplies from one end of Germany to the other. The autobahns were much admired in this country, mainly for the wrong reason (i.e. their original military function was overlooked). There was a clamour in the motoring press for similar roads in Britain; they would eventually be built as the motorways but not for thirty years – and after the world war, for which purpose the autobahns had primarily been built.

The only new roads in Britain in the 1930s were, as had been said, the urban 'bypasses'; although the length was probably less than 100 miles (160km) in total, they became something of a motoring icon in the later 1930s. Two or three around London had 'roadhouses' built beside them, the best known being the Spider's Web, the Ace of Spades

Opposite The Kingston bypass was one of a number constructed in the 1930s to relieve the growing traffic congestion in the towns. They were only partially successful as the lack of planning before the war allowed unrestricted building along the new roads known as 'ribbon development'. This photograph, though taken in 1951, shows the dual carriageway and the concrete surface of a typical prewar urban 'bypass' – as well as the extent of unplanned building alongside the road.

The world's
first choice
OPEN
MU 9396

and the Cartwheel. These establishments, complete with open-air swimming pools, bars and restaurants, catered for a new generation of young car owners driving out to them from London with their girlfriends in MGs, Lagondas, SS 100 Jaguars, Frazer-Nashes and other sports cars, for a carefree evening of dining, drinking and dancing to a popular dance band. They acquired a somewhat notorious, raffish reputation, though one supposes that to the present generation they would seem stultifyingly bland and boring! The construction of the bypasses, the roadhouses, the MGs and so on, were about to vanish for six long years when, in September 1939, the Second World War broke out. It was profoundly to affect British road builders and their plant in ways few could have foretold.

It had been known for some time that the Americans were very far in advance of the UK in the matter of road building and the plant they used. It was to the United States that the British government turned to supply the very large amount of plant and earth-moving machinery it was to require to build hundreds of airfields for both the RAF and, from 1942, the Eighth United States Army Air Force (USAAF).

Before the outbreak of the war, and for the first year, the airfields of the RAF were for the most past grass covered. For the airmen this was a very satisfactory arrangement for it meant that without fixed concrete runways, aircraft could always take off and land directly into wind, thus avoiding tricky crosswind landings. For small fighters, the Spitfires and the Hurricanes, grass fields were satisfactory and continued in use with the addition of a concrete perimeter track circling the field. Bombers were different

A section of the long overdue 'M' class motorways begun in the 1960s. Today there are just over 2,000 miles of motorways (like the M25 shown here) and despite the lurid reports of multiple pile-ups in foggy weather, they remain statistically the safest roads in the country.

proposition once the RAF had adopted the four-engined 'heavies': the Stirling, Halifax and Lancaster. These aircraft had to have a runway of a minimum length of 2,000yd (1,828m) of smooth tarmac or concrete runway. Because the wind direction was changeable the airfields had three intersecting runways, aligned with the prevailing and secondary wind directions. The runways were connected by a 50ft (15m) wide encircling perimeter track together with acres of hard standing and dispersal points. Of the 700 military airfields in use during the Second World War in Britain, 500 were built between 1941 and 1943. It was the biggest single civil-engineering project ever undertaken in this country since the building of the railways at the end of the nineteenth century.

The peak year of airfield construction was 1942, when the Eighth USAAF were in Britain in force: 60 heavy bomber groups equipped with B17 Flying Fortresses and B24 Liberators: four-engined bombers, 15 groups of medium bombers and 25 fighter groups. The building of airfields for the Americans alone amounted to starting a new construction on a 'green-field' site every three days. British contractors, using American-built heavy plant, constructed many and the US army engineers built at least ten major bomber airfields themselves. The statistics of airfield construction for the USAAF are impressive: the concrete laid for the runways, perimeter tracks and hard standing amounted to the equivalent of no fewer than 2,000 miles (3,200km) of six-lane motorways. A single heavy-bomber, Class A airfield required 175,000cu yd (133,875cu m) of concrete, 32,000sq yd (26,752sq m) of tarmac, the removal of about 8 miles (12.8km) hedges, the felling of over 1,000 trees, the construction of 8 miles (12.8km) of roads and the laying of 4½ million bricks – all that for a single airfield out of the 500 built during the war years. It was the methods and the plant used by the American engineers in airfield construction that was profoundly to change the perception of road construction in Britain after the war. It might, however, be pointed out that the wartime airfields were produced under the imperatives of war, without any regard for commercial considerations. Each Class A base cost, on average, £1 million (in the money of 1942).

The American heavy plant, introduced into this country for wartime airfield construction was much admired by British airfield construction companies, the most impressive being the Caterpillar D7 bulldozer, a machine virtually unknown to the prewar British construction industry. This powerful diesel-engined machine ran, as the name implies, on wide crawler tracks with a heavy curved blade at the front which was used to gouge huge amounts of earth to level as site. The wartime bulldozers were an early type; the blade was raised by a system of pulleys and was lowered simply by gravity, which meant that the blade did not dig into the surface to any great extent. Later models had full hydraulic rams which raised and lowered the blade to dig it in if required. The American forces used bulldozers worldwide; in the Pacific they were driven from landing craft as soon as the US marines had secured a lodgement in order to carve out an aircraft landing strip, often from virgin bush or a coral atoll. Armed men had shotguns to protect the drivers from attack from enemy snipers.

Other earth-moving and grading machinery which arrived with the Americans included the extraordinary 'Galion Grader'. These machines have been likened to a praying mantis; the grader had a blade set at an angle and a very long extension of the chassis which carried the front wheels. It was not a lightweight bulldozer, but a surface grader. The blade was set at a predetermined height and, as the machine was driven, skimmed the surface of loose earth. Since the blade was at an angle, any surplus earth was pushed aside to be levelled on the next pass or by another machine working slightly behind the first, leaving a level surface ready to be rolled into a road or aircraft runway. On a forward airstrip assignment the American engineers might use half-a-dozen graders in echelon levelling a long wide runway at a single pass. The RAF received some of these graders and a member of the wartime 5202 Plant Squadron, ex-corporal Maurice Gowlett, drove one just after D-Day in 1944; he recently had the opportunity of driving a preserved Galion and remembered that:

> I was just 21 when we landed on one of these in the afternoon . . . we disembarked from a tank landing-craft and drove to a collection point; just a collection of half-a-dozen fields . . . they were bulldozed into one and we created an airstrip for Spitfires to land just after D-Day.

It is difficult to see how that kind of instant forward airstrip, so vital to provide air cover to the armies on the ground, could have been created without the American plant that had been shipped over for the airfields in 1942 and the D-Day invasion in 1944; there was certainly nothing then produced in Britain which could have replaced it or many of the items of heavy plant the Americans used. There was, for example, the curious 'wobbly-wheel roller', as it was called by the soldiers who operated the device. This was light but not self-powered; the light weight was for transportation purpose only, for it had a very large tank which could take a considerable quantity of water ballast. Below the tank there were up to thirteen very fat-tyred wheels on eccentric axles. The wheels could also pivot from side to side – hence the nickname. The tracks overlapped and the result, when towed by tank or tractor, was a smooth runway rolled out of a field. In France the wobblies were used just after D-Day and they were also used in the sands of the Western Desert during the Eighth Army campaign in North Africa. The original wobblies were made in the USA in large numbers; others were made in England by Pullen Engineering, who continued to make them after the war.

Another American plant item that was imported during the war, the Barber-Greene finisher, looked at first glance like a mobile skip, but appearances can be deceptive and the Barber-Greene finisher was a formidably efficient machine that could, not-stop, lay a long road or airstrip. Hot asphalt was poured into a hopper from a tipping lorry. Two large Archimedes' screws fed the asphalt to a jigger which levelled and tamped it before it was fed to the 'screed', which was a large flat plate the width of the section being surfaced. From there a flat, hot carpet of asphalt was laid on to the road bed. This was a continuous process until the asphalt hopper was empty. After the war the Barber-Greene and other similar continuous finishers were to transform urban road building and resurfacing in Britain.

Not all the plant used in the great airfield building was from the United States. The humble British steamroller had a part to play; a small number were the sole means of levelling the hastily filled craters caused by the constant German bombing of the main RAF fighter airfield at Takali in Malta during the desperate defence of the island in the summer of 1942. As soon as the first raid of the day was over, soldiers would fill the craters and two or three steamrollers, kept in steam night and day, would roll the repaired sections of the runways flat, then retreat to revetments to await the next raid. Usually they did not have to wait very and, sad to say, all were eventually destroyed by enemy action, possibly the only steamrollers to be lost on active service. However, Malta was held.

In Britain, no new steamrollers were built during the war years but diesel rollers were constructed. The Aveling & Barford Company, after a good deal of tedious negotiation, were contracted to build a number of wartime diesel rollers to a specification issued by no less an authority than the Air Ministry Directorate-General of Works (AMDGW). It was to be a 'utility' job, expendable and, above all, lightly built to conserve steel. The twin tandem rollers, for example, were filled with concrete to compensate for the lack of metal. Avelings were appalled; they had always designed and built to far higher standards. Therefore, to protect their reputation and, one suspects, postwar sales, they had all the castings – engine, gearbox, etc. – embossed with a bold apologia that read 'War Design'.

After the Second World War the British plant industry clearly reflected the American wartime influence. The Romans had once shown what a road was; the Americans now showed how to build a road or runway quickly with efficient plant. Much of the plant sent to Britain during the war was either Lend-Lease or issued to the US forces; some was sold as war surplus at the end of hostilities and snapped up by contractors. British firms produced similar plant, the Vickers VR 180 Vigor of the mid-1950s, for example. This was a home-produced bulldozer which owed its running gear to the tanks that the company had been producing. It had a Rolls-Royce tank engine and was fast. Fred Knights, another ex-serviceman who collects classic plant, owns a Vickers VR 180 Vigor that he considers was in some ways superior to the American Caterpillar D8 bulldozer he also owns:

> . . . it goes without saying that the Caterpillar machines had quality and were durable, but they were slow; if you needed to speed up you had to stop to change gear. With the coming of the Vickers machine, it was much faster and you could change gear on the move. The track layout was different to the Caterpillar. The Vickers had independent bogies which gave a much smoother ride.

Despite the advances it seems the Vickers bulldozer faded out after about ten years. There were certain difficulties with the transmission, but Fred Knight thinks that they could have been eradicated. Vickers VR 180 bulldozers did work on Britain's first motorway, the M1, but then so did some veteran steamrollers.

With the construction of 2,004 miles (3,206km) of British motorways, the plant business has come of age. The civil engineering of the motorways stands comparison with any similar roads in other countries. The plant developed and used in the construction are now classic: from the surviving steamrollers to the newer diesel rollers, the scrapers, bulldozers, dumper trucks, concrete mixers and ancillary plant. In 1938 there were just 125,000 miles (200,000km) of road in Britain; sixty years later that has more than doubled to 242,924 miles (388,678km) of roads of which 2,004 miles (3,206km) are to motorway standard. Every yard has been constructed or reconstructed at some time or other by classic plant.

FACT FILE 1
Vehicles on British roads in 1995
Private cars and taxis: 353,200
Buses: 4,700
Motor bikes: 4,100
Goods vehicles: 68,900
Bicycles (estimate): 4,500
Total: 435,400

FACT FILE 2
Road mileage as at April 1995
Motorways: 2,004
Trunk roads: 10,916
All roads in the UK: 242,925

CHAPTER THREE

One Man Went to Mow: The Rise of the Combine Harvester

All is safely gathered in
Ere the winter storms begin.
(18th-century hymn)

SINCE PEOPLE FIRST planted seeds, rather than relied on the haphazard sowing of nature, harvest has always been a time of anxiety. The harvest is the culmination of the agricultural year. From biblical times through the centuries by rite, ritual and religion, and with sacrifices, corn dollies, harvest church services and secular festivals, human beings have offered thanks to their gods for the deliverance of a good harvest and to appease them to smile upon the next.

In the Western world today, a failed harvest is a disaster only in the fiscal sense. Not all that long ago it was a question of survival. In the 1930s unknown thousands starved to death in the USSR through the ruthless pursuit of a flawed political policy of collectivization, which resulted in failed harvests on a national scale. Famine still haunts the Third World.

Assuming, and it can be a large assumption, that the harvest is good, the farmer has to make the very difficult decision of just when to gather in the crops. Before mechanization (and to a lesser extent even today) if a farmer told his friends in the village pub that he had decided to reap, say, his barley, there would have been immediate doubts raised. Perhaps it was too early, perhaps it was too late; this would be followed by a litany of advice, most of it conflicting. There is rarely the precise moment when a harvest might safely be gathered. The farmer has to balance several conflicting factors, the most important of which is the one over which he has no control whatever: the weather.

Because the weather is vital to the harvest and because, in countries such as Britain, it is capricious, the time taken for the reaping of the harvest had to be as short as possible to take advantage of the all-too-brief good weather. 'Make hay while the sun shines' was excellent advice. However, haymaking was a relatively straightforward affair. Up to around the turn of the century, reaping was essentially still an activity that was unchanged since biblical times. Men and women cut the crop with a sickle, by reaching and holding a sheaf of the hay and then cutting it through with a single stroke. It was a back-breaking, laborious and inefficient process.

The great prairie-like wheat fields of the American Mid West – the birthplace of the modern combine. Here, a team of modern International-Harvester combines are working together, having cut a vast swathe through the standing corn on each pass.

Two Buckinghamshire farm workers stacking sheaves into a stook in 1935. Neat rows of stooked corn were a common autumnal sight in Britain until the universal adoption of 'combines' in the postwar years. In the background, much of the crop can be seen already standing in stooks of ripening grain. When judged ready, the sheaves were collected to be made into ricks to await the steam-driven threshing machine.

From an unknown date, reapers began to use a simple wooden crook, often a conveniently shaped tree branch, called in some English counties a fagging crook, to draw the standing hay sheaf together. It was then cut through with a broad-bladed sickle, the fagging hook (hence the saying, 'By hook or by crook'). After the hay was cut it was raked and left lying in the meadows for three days or so to dry. If the weather was good, and the hay dried, the sheaves were collected together by wooden rakes and pitchforked into the long, high-sided, four-wheeled horse-drawn wagons, the haywains, and taken to the farmyards to be made into hayricks. These were used by the farmer as winter feed for his animals. The time-honoured manual methods of haymaking, however picturesque, were very inefficient. Even when everything went well 30 per cent of the hay would be lost; if things had gone badly, up to 50 per cent, or half the crop, was wasted.

Traditionally in English counties haymaking time was around the second week in July. When the haymaking was over the next crop was the main harvest: corn. This was a far more protracted and labour-intensive proposition. It could mean, even in good weather, three or four weeks of dawn-to-dusk, back-breaking work for every able-bodied man, woman and child on the farm, aided by men lent from neighbours and itinerant workers, often Irish, who travelled from farm to farm, sleeping rough in barns and stables and who sought casual employment at harvest time. They usually found it, for a typical field of corn could have over a hundred people working on it at one time.

The work was divided into three distinct phases: cutting, stooking and threshing. The reaping or cutting of the standing corn was, dependent on the weather, followed by the binding together of the cut sheaf into sheaves and placing the sheaves on end into stooks – once a familiar autumnal sight. This was essential to allow the corn to ripen. The corn had to be cut before it was fully ripe, otherwise the grain would be scattered at the first touch of the sickle or scythe.

The second stage was also dependent on the weather as the stooks remained standing in the fields until they were considered sufficiently ripe. In most English counties oats would be left to weather for 'three Sundays', wheat for a shorter time of a week or two. The stooks were then collected, pitched with forks on to horse-drawn carts and taken to be stacked into ricks in the stackyard, perhaps as many as twenty on a medium-sized farm. Often a thatcher was employed to thatch the ricks, which were left to await the final phase of the harvest: the threshing to release the grain from the straw and chaff. Even the thatched ricks standing in the stackyards were at some risk: if the sheaves had too much moisture in them as a consequence of a wet harvest there was a very real danger of the whole rick catching fire due to spontaneous combustion.

Before the threshing, after all the stooks had been collected, the gleaners, usually the women of the farm, went into the stubble to 'glean' or collect fallen grain. After the gleaners had picked up all they could find – as with haymaking, a substantial percentage of the corn would be irrecoverable – livestock were let into the fields to forage for the lost corn.

Threshing was the process by which the grain was separated from the chaff and straw. Before the use of steam-driven threshing machines, threshing was at first done by tethering sharp-hooved animals and making them walk round and round over the scattered corn stalks to trample the grain from the straw. The animal system was slow and very inefficient. Later, threshing was done by a group of men and women using flails and wind winnowing. Although an improvement on animal threshing, it remained such hard manual work that it was often left for the fallow winter months.

Winnowing goes back to biblical times. The threshed corn and chaff is placed in flat baskets and thrown into the air; the wind blows away the light chaff, leaving the heavy grain to fall into a pile at the feet of the winnower. Barns were built with doors at each end, and were aligned, if possible, with the prevailing winds to allow a draught to pass through the building for the very purpose of winnowing the grain in the winter months. Few agricultural workers today would be prepared to stand for hours in a freezing winter's wind continuously blowing through the unheated barn as they winnowed the grain.

Threshing and winnowing the entire harvest could well involve daily work for a number of men throughout the winter months. By about 1840, after many false dawns, the first practical engines as aids to harvesting were in fairly widespread use in Britain. There were several types of 'fanning mills' that separated the corn from the straw. The sheaves were fed into the machine to either beaters or 'scutching' rollers, usually three, which were in effect rotary rakes that separated the grain from the straw. The heavy grain fell through sieves, leaving the straw to be fed to a chute at the side of the machine. As the grain fell a fan blew the chaff from it.

Although both horses and water-wheels were in use to drive the early threshing machines, a typical, small manual machine required three people to operate it: one to turn the heavy handle which drove the machine, one to feed in the corn sheaves and a third to hold a sack beneath the mill hopper to collect the separated grain. It was more efficient, quicker and independent of wind for winnowing, and less drudgery than the flails. However, the fanning mills still remained very hard work indeed.

Though threshing was hard, unremitting work, by far the most labour-intensive stage of the entire harvest was the first one, which was the cutting of the standing corn. By the eighteenth century the scythe was in widespread use, but many reapers on smaller farms still preferred, right up to living memory, the older 'hook and crook' sickle. The scythe seems, to a layman, to be far more efficient – one sweeping pass and a great deal of corn would be cut. This was true, but there was a difficulty. When corn is cut with a sickle and hook all the heads would be the same way up; with a scythe the corn would topple either way and a second man would have to rearrange the heads all the same way when he picked up the sheaf to bind it. This was a time-consuming process. At some time in the sixteenth century, in order to overcome this problem, the Quakers produced scythes that were fitted with a light wooden frame made from willow attached to the shaft of the scythe and blade. This was called a creet (from the Old English word *crete*, a baby's cradle). As the corn was cut by the sweep of the scythe, the cut stalks fell on to the creet. All the heads were the same way up, and were carried on the creet until the scythe reversed for the next pass. As it did so, the cut sheaf fell from the creet with the corn heads still together. A second man or a boy then collected the sheaf, and twisted a bond around to tie it. The binding was never string but specially made up from corn stalks. The tied sheaves were then stacked into the stooks.

Even after the arrival, towards the end of the nineteenth century, of the steam-powered threshing machine, the cutting was still done by hand with sickles or scythes. With a scythe fitted with a creet, a man might cut about 4 acres (1.6ha) of wheat in a long working day. In practice it was usual for several men to work together as a team. True to established country custom, they would elect a leader (in Suffolk he was called the 'king of mowers'), who was usually the strongest and most experienced man in the team. It was he who set the pace as the mowers slowly advanced abreast across the cornfield all swinging together like a rowing eight, the multiple steel blades ringing as they scythed through the tough corn stalks. As they did so, the blades needed to be sharpened at fairly frequent intervals if they were to cut cleanly. The reapers each carried, in a leather holster, whetstones about 10in (25cm) long, known as strickles to touch up the cutting edges of the long blades of the scythes.

The reaping would continue throughout the daylight hours. The men would get an hour for their lunch; then, laying their scythes against a hedge, they would sit on the headland under a shady tree for their 'bait', or midday meal, which their wives had provided. It was usually bread, cheese and pickle, if lucky a slice of cold game or rabbit pie, with a stone flagon or two of strong farmhouse cider or beer provided by the farmer. Then it was back to work until it was either dark or the whole 20–30-acre (8–12ha) field was cut. The very last sheaf was traditionally made into a corn dolly, a crude human figure said to bring good luck to the farm. This was a custom surviving from pre-Christian, pagan beliefs.

When the grain held in the stooks was considered sufficiently ripe and the whole harvest had been safely gathered in, there would be celebrations. The last wagon of corn sheaves would be led off the field with all the harvesters gathered round it, laughing, joking and possibly even singing local harvest folk songs. They all looked forward to the traditional harvest supper provided by the farmer. Sadly, this tradition had begun to die out before the First World War. In place of the supper, the gathering in of a successful harvest was marked by an extra payment of 2s 6d (12 ½p) each for the farm workers – serious drinking money in those far-off days, to be followed by a strictly sober, church-organized, harvest home festival. Harvest festival or not, the idyl-

A team of reapers towards the end of the 19th century. Their scythes have 'creets' fitted to collect the cut grain. In the background, the farmer's wife brings the reapers their 'bait', the welcome midday meal. It is interesting to see in this delightful 1885 Seguin engraving that thistles and other weeds flourished in the corn fields of that herbicide-free age.

lic scenes of reapers scything through a field of corn, much admired by the makers of 1930s film documentaries, was labour on a scale that today would be considered totally unacceptable both in human and economic terms.

The first successful steps towards the mechanization of the reaping process was made not, as might be supposed, by an agricultural engineer but by a man of the cloth. He was the Reverend Patrick Bell, who produced, in conditions of some secrecy, the first practical horse-drawn mechanical reaper in 1828. If the Reverend Bell had read Classics instead of Divinity he might have been accused of reinventing the reaper, for Pliny and Palladius both make mention of a cart or *vallus* which, pushed by a mule or horse into standing corn, cut or stripped the heads off the grain. It would seem that the device was used in the first century AD in southern Gaul. Just how it worked is unclear. It must have been a fairly moderate performer for it was to be supplanted for nearly 2,000 years in favour of sickle and scythe reaping, until Patrick Bell and others turned their attention to the problem. Bell, Pliny aside, was far from the first to propose a mechanical reaper. There are many claimants prior to him, but Bell's machine was the first to demonstrate its ability to cut the standing corn and to lay the cut sheaf the right way round on a board.

Bell, a Scot, was so secretive about his work that all his early experiments were conducted in a closed barn assisted by his brother with, unbelievably, rows of hand-planted corn on which to experiment. Secretive or not, Bell's machine was, in its essentials, the true prototype of the modern reaper, many of the features being found in later power-driven reaper-binders.

Bell's reaper had cutters which were exactly like those of a modern reaper or a DIY hedge trimmer: triangular fingers with two ground cutting edges screwed on to an upper and lower horizontal bar the length of the intended swathe. The two bars were in rubbing contact, the 'fingers' acting as multiple scissors as the upper bar was given a reciprocating motion. In effect, this opened the scissors, cleanly cutting the stalks. Since the points of the cutters were like a comb facing the standing corn, the stalks entered the 'V' of the cutters to be held until cut. They fell backwards on to a sloping board or canvas, sliding to the ground as the reaper advanced.

Bell, secretive as ever, insisted that the first field trial of the reaper was made at night. The trial was considered a success and the machine was demonstrated publicly at a local farm called Greystones. The reaper was mounted on two large cartwheels and pushed by two horses, the driver walking behind them. This rather unusual arrangement was presumably to avoid the necessity of the horses trampling on the uncut grain. One of the features of the Bell machine was the now familiar rotating wooden 'sails', which drew the standing corn towards the cutters and also placed the cut sheaf on to the board to fall to the ground in order to be collected and tied into sheaves by a second man walking alongside. In trials the Bell reaper could cut about half an acre (0.2ha) an hour or 6 acres (2.4ha) per 12-hour day. A man with a scythe could only manage 1 acre (0.4ha).

There were, of course, snags. The alignment of the cutters was a constant source of concern; if only one was out of alignment it affected the performance of the whole cutting sequence. The cutters had to be dismantled and reground after every 50 acres (20ha) or so. The whole of the mechanical process was driven by gearing from the land-wheels; in the days before mass-produced precision ball-bearings the frictional losses must have been high, placing a heavy burden on the two horses, even on level ground.

After the trials at Greystones, the reaper was exhibited at Highland shows and several production versions made. These were not all by Bell, who had declined to patent

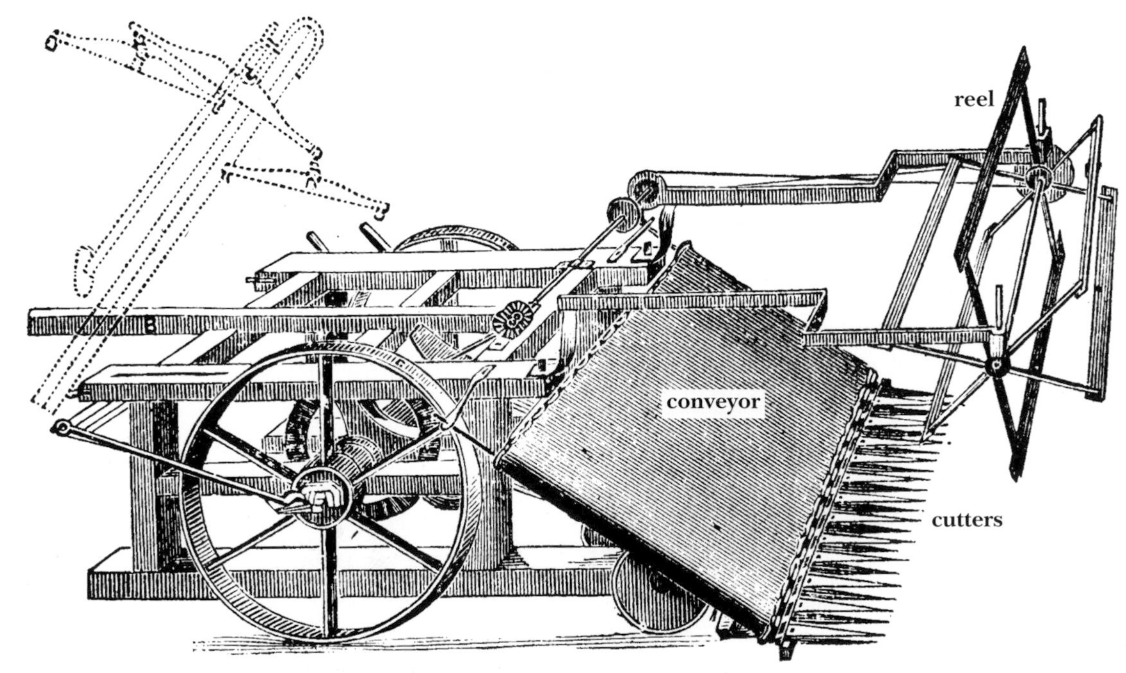

Bell's Reaper can lay claim to be the first step towards the combine harvester. The essential features which Bell invented are still valid: the cutters, reel and the conveyor for the cut sheaves. Only the rasp bar and straw walkers were yet to be introduced. As a reaper, Bell's machine worked well given the limitations of the method of propulsion and powering. With just two horses, pushing it must have been both under-powered and very difficult to steer.

his reaper, believing that he had been sufficiently rewarded by the successful outcome of the trials and wishing, altruistically, to see it benefit mankind. In fact, the Bell reaper was never to receive the commercial success it merited. By the time it was shown, in the Great Exhibition of 1851 at Crystal Palace, London, several rival American-made reapers were also on show and available. (There are claims that one of Bell's machines had been exported to America and become the prototype for these American reapers.) It is beyond doubt that at the Great Fair in New York, also held in 1851, the firms of McCormick and Hussey had exhibited their reaping machines.

In 1831, two years after Bell's machine had been demonstrated, Cyrus Hall McCormick, a Virginia farmer's son, built his first reaper. It was tried out by cutting oats on a farm which adjoined his father's. He, unlike Bell, lost no time in applying for a patent. An improved machine was soon in commercial production.

The McCormick reaper was essentially the same in principle as Bell's in the use of oscillating scissor cutters with the same rotating wooden sails to feed the crop to the cutters. One major difference, however, was that unlike Bell's machine, the McCormick was pulled by a single horse, rather than pushed by two – horses pull better than they push – and the cutter comb and rotating sails were offset to the left. This meant that when starting on the headland to cut the first swathe of the crop the horse, on subsequent passes, was always treading in stubble and not through uncut crops.

The driver of the McCormick reaper sat on the machine, with a second man walking beside it, who raked the cut sheaves from a wooden platform. McCormick then added a seat to the platform for the raker, and the performance improved to some 8 acres (3.2ha) a day. A third version of the reaper dispensed with the second man, the raker, altogether. A mechanical pusher geared to the cutters cleared the cut sheaf from the platform so that it lay on the ground at regular intervals to be tied, and placed in stooks, usually by the women of the farm.

By 1847 McCormick had competition from another maker, Hussey. Both firms set up reaper factories in Chicago, which had become the centre of the American corn trade. The factories there were producing, by 1853, some 500 mechanical reapers a year.

The almost instant success of the American makers was in contrast to the apathy shown to Bell. The reason is not far to seek. The average British cereal field was around 25 acres (10ha). In the American Midwest and Canada, 1,000-acre (400-ha) prairies and more were becoming commonplace, as the plains were put to the plough. To reap 1,000 acres (400ha) of wheat with scythes would be like trying to mow a football pitch with nail scissors. American farmers *had* to have machinery as the population, fuelled by successive waves of immigration, grew. This created increasing urbanization which, in turn, drew farm workers away from the land to far better-paid jobs, under far better working conditions, in the burgeoning cities. Thus, at a time when the demand for cereals was increasing year by year, there were fewer men to work the fields by the old, labour-intensive methods. American agricultural machinery-makers found, therefore, a ready market for their new products, mainly at home, though a number were also made available for export.

The importation of American reapers into Britain did, to some extent, stimulate the home market. Bell's reapers, in an improved form, were reintroduced and in 1860 the British firm of Samuelson and Company of Banbury, Oxford, produced the American-designed Dorsey mower. After the enforced interregnum of the American Civil War,

The sheaves being collected in the picture on page 56 could well have been cut by this horse-drawn reaper/binder. The year is 1935 and rural England of that time is well depicted: the horses plodding chest-high through the standing corn, and the sheaves dotting the undulating field in the wake of the binder, with orderly rows of stooked corn in the distance.

progress in the United States was rapid. In 1865 the American Marsh reaper featured mechanical bunching of the sheaves by the use of a canvas conveyor; the sheaves were then tied and unloaded by two men aboard the reaper, leaving a trail of ready-to-stack sheaves.

The eventual goal of the combine harvester – though no one at that time could have been aware of the fact – was brought a step nearer when the first reaper-binder appeared. This was the machine made by Walter A. Wood in 1873, which automatically tied the cut sheaves by twisting a wire binding around them. It worked well enough, but farmers had objected that the wire binding, cut and discarded in the straw when the sheaves were threshed, could injure cattle.

The man who had introduced automatic wire-binding was John Appleby, an Englishman working in America. After the rejection of wire-binding, he set to work to devise the much more difficult task of binding the sheaves with twine and, by 1878, had perfected an ingenious mechanical knotter which tied the sheaves with harmless string. His 'knotter' was to remain in universal use for the next eighty years.

The true reaper-binder was now a reality. McCormick astutely obtained the rights to Appleby's knotter and by 1884 had sold no fewer than 15,000 reaper-binders which featured it. By the 1880s in Britain, at the agricultural shows around the country, a large selection of reaper-binders was on offer, most either American or made in this country under licence from American companies. At one show, the 'Royal' held in Kilburn – which to Londoners today must seem a most improbable venue – Bell's original reaper was shown, one that had been demonstrated in Scotland in 1829. It would, therefore, have a strong claim to have been the first reaper to work and the first to acquire classic status. There is the question as to whether Bell's machine had influenced the American Cyrus Hall McCormick. The 1884 show catalogue stated of Bell's original reaper that: ' . . . a number of these machines had been sent to America in 1834'. The date of 1834 was four years after McCormick had built his first reaper but it is possible that, as noted, a single Bell machine might have been exported to the United States earlier.

By the late 1880s, the horse-drawn reaper-binder had reached a plateau in development, the limitation being the load two or three horses could draw. The problem was simply that, as more functions were added to the machines, the power requirements grew too, which could only be provided by gearing or chains from the land-wheels. In his informative book, *Country Voices,* Charles Knightly interviewed several old farm workers, including Bill Denby, who had worked on the horse-drawn reaper-binders which were still to be found lingering on small farms up to the late 1940s:

> . . . we generally had four horses . . . if you only had three though you had to keep resting them or they would soon tire out. It was a very tiring job for the horses because [the] machines were all ground drive. All the moving parts – the sails, the knives . . . were driven by chains from a ground wheel and of course the horses had to pull that along: [the wheel] had spade-loops [spuds] on, so it wouldn't skid, but it did skid, and if there'd been a lot of wet it didn't use to drive well.

By the 1870s, it was clear that any significant reaper-binder developments were dependent on a source of power other than that of horses. Steam traction-engines had been available from about 1860, but at around 10 tons they were too heavy to be considered for field use. However, at a Royal Show held in Birmingham in 1875, a McCormick reaper-binder was shown, pushed by a small 8nhp (nominal horsepower) Aveling and Porter steam traction-engine. The combination was demonstrated at Leamington, reaping a 12ft (3.6m) swathe so perfectly as to 'surprise the spectators and delight all'. A gold medal was awarded. But, for use on the average British farm, the same objections that had ruled out the steam traction-engine for direct ploughing still applied: it was too heavy and too expensive.

Static steam engines were, however, in demand for threshing. These were small portable ones at first; they were not self-propelled but were drawn to the farm by horses, and were owned by contractors who would undertake to thresh all the crop. To

A typical steam threshing scene. The engraving is an accurate representation of a once familiar scene: the traction engine with driver in attendance, two men feeding the sheaves into the thresher with its multiplicity of external belts and cogs, and two more bagging the threshed grain. The alignment of engine and threshing machine was critical if the driving belt was to remain in place.

that end, contractors would undertake to supply, on hire, all the machinery required for threshing: the steam engine, the threshing machine, elevators to lift the sheaves and, later, the straw, and a chaff-cutter. All had to be collected from the contractor by the farmer, requiring no fewer than twelve horses. The portable engine alone needed four horses to pull it, four more for the threshing machine, and two each for the elevator and the chaff-cutter. It was usual for farmers to borrow horses from a neighbour who, later, would probably take over the machinery for his own threshing. By the turn of the century, the portable engine had given way to the self-propelled traction-engine which replaced the horses altogether, and towed the whole threshing kit in a four-vehicle coupled convoy, through the villages and lanes, often in darkness, to get to the farm and be ready to start work at first light the next day.

Once the contractors arrived at the farm, a level stretch of farmyard would be selected, ideally next to the first rick to be threshed. The threshing machine, called a drum or a threshing box, was levelled and its wheels securely chocked. The traction-engine would be carefully aligned with it; this was vital as the long fabric drive-belts, running over the flywheel of the engine, and the driving wheel on the thresher, had to be exactly in line because neither wheel had flanges. That the loosely tensioned long belt stayed on the plain rims was (and remains, when steam threshing is demonstrated at a present-day steam rally) a source of wonder.

In the many unpaved farmyards of the time, the two contractor's men often had great difficulty in setting up the threshing machine. It weighed up to 3 tons and had to be set absolutely level; a spirit level was used in conjunction with jacks and wooden packing under the wheels to achieve this. The heavy traction-engine could also pose problems; most farmyards were a sea of mud in winter when the threshing took place. In *Country Voices* Will Flinton describes how an engine sank up to the flywheel in one particularly boggy farmyard. They had to send for another engine to drag it out with a long steel cable.

Steam threshing began in September, as soon as the farmers considered that the stacked corn was ready, and continued from farm to farm until May. Fortunately for us, the height of Victorian steam threshing coincided with the advent of photography, and so contemporary scenes of threshing have come down to us. The engine was always immaculate with the brass lagging bands around the boiler highly polished and its paintwork (green, red or brown) gleaming, as was the steel of the motion. The driver would stand on the engine, which rocked gently as it worked. He kept an eye on the water gauge and stoked a bright fire to keep the steam pressure up to the mark at about 120psi (8.4kg/cm^2). One such driver was the ex-shepherd, Will Flinton, mentioned above. Born in 1900, he has left an account of his work in the 1920s:

> There'd be so many men on the stack [rick], putting sheaves down on to the drum, and there would be a man stacking the straw at yon end o' the drum, and another carrying corn at this end. They would all be farm chaps, but one of us [contractor's men] would feed the sheaves into the drum: he had what we call a 'band-cutting knife' in his hand, and he used to cut the band [binding string], round the sheaf and shed it all into the threshing box. One of us would do that, and another would drive the engine, and we changed over every two hours.

Midday meal break for a Berkshire threshing crew in the 1930s. The traction engine simmers with dampers shut. That the men have worked hard is evident from the bagged grain. Now they and their coated foreman relax, eating bread and cheese with a gallon or two of strong cider in the wickerwork-covered flask to wash it down; then back to work until the light fades.

The farmer would be responsible for the provision of water and coal for the engine. This was good hard Welsh steam coal, which in 1914 cost about one shilling (5p) a hundredweight (50kg), brought by wagon from the nearest coal merchant who could then be found at any railway station. The engine, when threshing, would use 5cwt (250kg) per working day, a total outlay of 21 shillings (£1.05). The other expenses were not high either. Will Flinton recalls again in *Country Voices*:

> . . . we used to charge 45 shillings a day (£2.50), for however many days we threshed . . . and there was two men to pay out of that, and there was all the tackle to keep in first-class condition, . . . and then you'd have all the oil to get, cylinder and lubricating oil for all the bearings: and all the little belts on the threshing box and the elevator, they all had to come out of that same money . . . my [weekly] wages could be 30 shillings (£1.50). But if it was a wet day and you couldn't thresh, we were only getting half a crown (12½p) a day: three wet days in a week you wouldn't be taking home anything like 30 shillings. In summer we often had nothing to do at all.

It is pleasing to be able to report that Will Flinton managed for find use for his skills by driving a County Council steamroller as his summer job.

Once the threshing machines had all been set up, they might remain at a given farm for a week or more to thresh all the grain in the ricks. The two-man contracting crew would cycle home each night and face the fact that they would have to get up at five in the morning and cycle back to the farm to light up the engine so as to have a working pressure by seven o'clock. The threshing machines used for steam threshing from the 1880s to the late 1940s were a development of the early, large machines that were driven by horses tethered to walk in a circle. Steam-driven threshing machines were made by several agricultural engineers, Ransome & Sims, Barrett & Exall, Clayton & Shuttleworth and Garrett & Hornsby being some of the best known. As an example of the rapid expansion of the steam-driven threshing machine, one of the firms, Clayton & Shuttleworth, were selling a hundred machines a year by 1891.

The steam-driven threshing machines had, in essential, been stabilized by 1890 and would remain, at least in Britain, more or less static until the establishment of the modern combine harvester. British threshing machines were made of wood and had a proliferation of external belts and chains. The sheaves, after the binding was cut, were fed into the top of the machine to rotating drums with tines attached to them to carry the sheaves down into the threshing cylinders. This operation was extremely dangerous and many horrific accidents occurred because of careless loading. In 1886 Ransomes introduced an enclosed safety self-feeder, though it was not universally adopted and feeding accidents continued into the 1950s.

Early steam-driven engines threshed the sheaves by peg rollers but these tended to damage the straw. Typical later machines, made by Ransomes or Marsh, used the rasp-bar system, the elements of which remain in many combine harvesters to this day. There were many variations, but in simplified terms, the sheaves were fed to a rotating threshing cylinder which had a series of serrated rasp bars set rather like the blades of a cylinder lawnmower. The rotating rasp cylinder had, around the lower half, a closely fitting steel 'concave', which had vertical perforations over the entire surface. The clearance between the concave and the rasp bars was adjustable to fine limits, which allowed the straw to pass through without damage but stripped the grain from the heads. As the sheaves were fed into the threshing cylinder, grain separated from the straw by the rasps fell through the mesh of the concave. The straw, chaff and any remaining grain were passed to a secondary beater that moved the straw to 'straw walkers'. These were oscillating, serrated, wooden arms which shook any remaining grain from the heads and drew the straw forward to be ejected from the rear of the machine. The separated grain and chaff dropped through a series of shaker riddles and sieves to be dressed.

The final section was equipped with a powerful fan. It was a form of wind winnower; the fan blew the light chaff from the grain, which fell through the chaff spout out of the machine to the ground. A further series of graded riddles separated weed seeds, short

An early photograph of a steam threshing set with a portable steam engine, threshing machine, and an elevator for the straw being formed into a haystack for winter feed.

The nostalgic appeal of steam threshing is such that whole sets are preserved in working order, like this one threshing at the 1997 Blandford Steam Fair. Not only was it seen to be working, visitors to the Dorset fair could buy the flour freshly milled from the newly threshed grain. The enthusiasts who painstakingly restore traction engines and threshing machines – often salvaged from scrap – are perpetuating a recently vanished technology.

pieces of straw and other detritus from the grain to exit spouts under the thresher. The dressed grain passed to a hopper and then, via hand-operated gate, through the grain spout, which had hooks around the periphery to hold sacks while being filled with grain.

The third item of the threshing equipment was the straw elevator which, also driven from the traction-engine, raised the straw from the threshing machine whence it was pitch-forked on to an endless travelling band, like a modern escalator, the height of which was adjustable as the haystack, fed by the straw from the elevator, grew larger. Men with pitch-

Before the tractor became universal, natural horse power was used for such tasks as driving an elevator which lifted unthrashed sheaves to form stacks. Hard manual labour was still involved as every sheaf had to be lifted on to the elevator and then evenly spread on the growing rick, which would be finished off by thatching to keep it dry until threshed during the winter.

forks stood on the top of the stack to level the straw. The final, though optional machine, is self-explanatory: the chaff-cutter, into which the ejected piles of chaff were shovelled.

By 1900 more than three-quarters of the corn crop was being harvested by machines. It might be supposed that the farm workers, released from the drudgery of the scythe and flail, were grateful. This was, in fact, far from the case; it was not unknown for threshing machines and reaper-binders to be mysteriously wrecked. But progress was unstoppable.

Towards the end of the nineteenth century all the elements for a combination in a single unit of the existing reapers and threshers, a nascent combine harvester, were in place. The first machine was American. It was produced by Daniel Best of San Leandro, California, in about 1886. This machine was huge; it weighed 15 tons and required no fewer than forty horses to pull it, controlled by a single teamster. The Best machines were suited to the limitless prairies on which they were intended to operate. Though only powered from the land-wheels, on those firm, dry, level plains the massive machines did work. There were often two or three in echelon, reaping and threshing the standing corn, with each cutting a 35-ft (10.5-m) swathe. The processed grain poured out through a long delivery pipe into wagons drawn by more horses travelling alongside. They were magnificent, though hardly practical outside the prairies of the American Midwest. It would be nearly fifty years before the modern combine harvester appeared; nevertheless, Best's reaper-thresher had shown the way.

In 1910 the Canadian agricultural engineers, Massey-Harris, produced their No. 1 reaper-thresher. It was far smaller than the Best machine, but was still horse-drawn and the machinery was still driven by the land-wheels. The No. 1 was probably in the nature of an interim proposal and progress was slowed by the war of 1914–18, but

Early days of combining: August 1929 to be precise, when this Massey-Harris 'Harvester/thresher' No. 5, one of the first in Britain, was being evaluated on Mr Bragg's Hampshire farm. The man with the trilby is bagging the grain. The width of the cut is impressive, though the tractor must have had a hard time pulling the big machine (it certainly needed its 'spudded' rear wheels). Interestingly, the rest of the crop appears to have been 'stooked' conventionally.

Massey-Harris continued to develop the reaper-thresher and by 1922 announced the No. 5. This, for the first time, had a petrol engine to power the reaping and threshing machinery. It also offered a rasp-bar option for the threshing. With the land-wheels now only subjected to rolling loads and axle friction, the No. 5 would have been a good deal easier on the horses, though by 1922 petrol-engined tractors with sufficient power were becoming available to haul the Massey-Harris machine. The Massey-Harris No. 5 must be considered a classic machine.

The Best Company, which had produced the forty-horse machine had, perhaps as a consequence, amalgamated with the Holt Company. In 1922 this amalgamated company produced tracked petrol-engined tractors to haul reapers on the soft but fertile soil of the San Joachim Valley. The firm became the Caterpillar Tractor Company, disposing of the Holt-Best harvester interests to another tractor-maker, John Deere, in 1927. Track-laying 'caterpillar' tractors, developed for military use in France during the First World War, were found to be ideal for use on the vast American wheat fields hauling the new, relatively lighter, reaper-threshers like the Massey-Harris No. 5, the low footprint of the caterpillar-type tractor causing very little, if any, compacting of the soil.

There were now other reaper-threshing machines on the market. In 1930 McCormick joined the Dearing Machine Company to become the International-Harvester Company. This company produced a number of 12-ft (3.6-m) cutter machines with rasp-bar threshing, light enough to be drawn by low-powered wheeled farm tractors. Case produced the tractor-towed Model Q, also with a 12-ft (3.6-m) cutter, able to process 25 acres (10ha) of wheat into 30 tons of grain in a single day.

Massey-Harris seem to have been the pacemakers, for the Canadian company continued to develop their designs of what was still generally known as a reaper-thresher. By 1937 the company was offering the No. 15, a small (by previous standards) reaper-harvester, which was among the first to be powered by the tractor's rear-mounted power take-off. The linkage resembled the propeller shaft of a lorry, complete with universal joints at each end to permit turning.

With the smaller farmer in mind, Massey-Harris, in 1938, offered the 'Clipper', a 5-ft (1.5-m)-cut 'combined-harvester', at a cost that most farmers could afford. This machine was driven by tractor take-off. All the various makers of 'combines', as they were now generally known – Massey-Harris, International-Harvester, Allis-Chalmers, Oliver, Minneapolis Moline, Case, etc., although differing in size and detail, worked in much the same way. They were powered by tractor power take-off with scissor cutters fed by rotating sails, or 'reel', at the front and rasp-bar threshers, the dressed grain being fed from the delivery trunk to a truck running alongside or held in a hopper. The straw and chaff jettisoned was, usually, burned later. Most small to medium harvesters could be operated by one man: the tractor driver.

The next big development was to bring the combined-harvester into the modern era. It was the Massey-Harris No. 20. This 1938 machine was the first to be self-propelled, the intrepid driver sitting in the open at the front a little above the whirling sails and clattering cutter. The No. 20 had a 16-ft (4.8-m) cutter and was powered by a 65hp Chrysler petrol engine. After eight pre-production machines had been discreetly tested in Argentina, 925 were built in 1938 alone. This historic, classic design was a little on the top-heavy side and was supplanted in 1940 by the same firm's No. 21. Many

One of the red painted 500 Massey-Harris No. 21 combines which formed the 1944 'Harvest Brigade' in the United States. The self-propelled No. 21 combine-harvester was to revolutionise farming first in the United States then, via wartime Lend-Lease American machines (mainly M-H No. 21s), in this country, too.

of these machines were shipped to Britain to help in wartime food production, as the Fordson tractor had been in 1917. The contribution made by the No. 21 'Massey-Harris Harvest Brigade' to the American war effort in 1944 was indeed a wartime epic.

The 1942 American and Canadian cereal crop was the largest on record, the corn crop exceeding 3 billion bushels, the wheat a billion bushels and the Canadian wheat crop over half a million bushels. Even with this plenty, such was the wartime demand that rationing had to be introduced. Farmers were therefore encouraged to produce even more in 1943. They succeeded to the point where there were insufficient machines to harvest the immense crops. Combines were made of steel but so was practically everything else, particularly ships, guns and tanks. The combine-makers were strictly rationed. The grain silos and storage elevators were overflowing and still tens of thousands of acres of standing corn and wheat awaited harvesting.

Joe Tucker, vice-president and sales manager of Massey-Harris USA, approached the War Production Board and said: 'Allow us sufficient steel to build 500 combines over the present allocation and the self-propelled No. 21, with a single operator, would produce more grain for a given amount of steel than any other.' The idea was that the 500 machines would work on US farms that had no combines under the supervision of Massey-Harris men. This would release over 1,000 tractors for other work and thus save half a million gallons (2.3 million litres) of fuel. The target was to harvest 15 billion bushels of grain in a military-style operation. To achieve that target the 500 red-painted No. 21s would have to average no fewer than 2,000 acres (800ha) each.

The combines left the Toronto factory in March 1944 on thirty railway flat cars, each combine emblazoned with the words 'Massey-Harris Self-propelled Harvest Brigade'. Unloaded, they were transported on lorries to the farms (with those 16-ft (4.8-m) wide cutters!). The trail ended 1,500 miles (2,400km) from the starting point the following

September. They had made good Joe Tucker's promise: 1,019,500 acres (407,800ha) had been harvested of 25 million bushels of grain. The estimated savings were 300,000 man-hours and 500,000 gallons (2.3 million litres) of fuel. One Massey-Harris man, Wilf Phelps, who worked in Arizona, was the top operator. He had single-handedly cut 3,438 acres (1,375ha). He had proved that one man could indeed mow a meadow – a very large one at that. The operation was ostensibly done to aid the Allied war effort, and no doubt it did. It did not do Massey-Harris any harm, either. In 1947, in one small town in Kansas, over 2,000 Massey-Harris self-propelled combines were counted passing through in just five days.

One, perhaps unforeseen, effect that the combine harvester had on the tempo of harvesting was this: because the new machines processed the harvest as soon as it was reaped, the need for stooking and storage into ricks no longer applied. This meant that the grain had to be harvested when perfectly ripe, which in turn required the corn to be left standing, uncut, for longer. This made the farmer a hostage to fortune in countries such as Britain with unpredictable weather. However, the cost-effectiveness of 'combining' was such that the risk was deemed worth taking.

The story of the combine harvester is to a very large extent American. The reasons for this are the sheer size of the US grain farms, offering a viable market, and the imperative of war but without many of the collateral restrictions that were imposed on British manufacturing. Chronic shortages of materials and skilled manpower, a universal and strictly enforced blackout, and air raids by day and night all inhibited Britain's combine-makers when considering new, radical designs.

Lacking a home-produced combine in significant numbers, the British government, in the shape of the Ministry of Agriculture, sanctioned (under the terms of the Lend-Lease Act) the importation of American machines, mainly Massey-Harris No. 21s, or the Allis-Chalmers All Crop 60s. The performance of the wartime machines belied the prewar assertion that the climate in Britain was too damp for the combine.

The wartime American machines showed the way. Some British makers did attempt to upgrade their standard wooden threshing machines, which had been designed before the war, by the addition of an integral diesel engine to replace the rapidly vanishing steam traction-engines which, traditionally, had powered threshing machines on British farms. Fosters of Lincoln, for example, offered a diesel-powered machine in 1943. It was progress of a sort, but a long way from a true combine harvester.

One British company, Clayton & Shuttleworth of Lincoln, which had made threshing machines, did produce a horse- or tractor-drawn combine as early as 1929. However, almost immediately, the firm suffered financial difficulties and Marshalls of Cambridge took over the Clayton combine business, though the machines were still produced under the Clayton name. Altogether, 130 machines were built, mainly for export. The number known to have been sold to farmers in England was eight, and one more to Scotland. Not surprisingly, in view of the meagre home sales (a consequence of the depressed state of British agriculture during the 1930s) all production of the Clayton ceased in 1933. The prospect of competition from the first American combine imported into Britain, the aptly named International No. 8, may have hastened the end of the Clayton.

One Clayton, No. 116, survives. It is the 'Scottish' machine, later owned by Lord Balfour, who presented it to the National Museum of Scotland. It has been restored to working order by a team of enthusiasts of the Scottish Vintage Tractor and Engine Club in Perth.

It is thought that the number of combine harvesters in use on British farms in 1942 was 950. By 1944 that figure had risen to 2,500, nearly all the machines being American 'Lend-Lease' imports. Granted that food production was, during those years, on a wartime footing, nevertheless the figures showed that, given the opportunity, British farmers would (contrary to the prewar experience of pioneering makers like Clayton) gladly use combines if available at a reasonable cost. In short, the signs were that when the war ended, the market for home-produced combines would be bullish.

Bullish the potential market might have been but, even with the return of peace in 1945, British agricultural engineers were still handicapped by chronic shortages – notably of fuel – which would persist for some time. These were the years of austerity and the controls enforced by a tight-fisted chancellor, Sir Stafford Cripps. British makers were, therefore, slow to produce new combines that would compare with imported American-made machines. Progress, though slow, was made. Ransome of Ipswich, J. Mann of Bury St Edmunds, offered tractor-drawn combines. By 1950 David Brown, the tractor manufacturer, joined with McGregor Guest and offered a newly designed combine that bagged the grain.

But despite these and other exceptions, it was a story which, in the postwar years, was to become depressingly familiar. Big American companies, who had amortized their development costs in a huge home market, gained footholds in Europe by the simple expedient of assembling their machines, which were adapted for use on the generally smaller European farms, and they were able to offer them at highly competitive prices. International-Harvester, for example, made an excellent combine, the B64, in Doncaster, Yorkshire. The 'B' stood for 'Britain' to distinguish it from the US-type 64. In 1956, the American company John Deere bought into the German Lanz Company, which was then renamed John Deere Lanz and produced their Model 250 combines for the European market. It was not until 1958 that Ransome of Ipswich who, before the war, had been a major supplier of traditional threshing machines, driven by steam engines, made their first diesel-engined self-propelled combine, the 902. This machine could combine 3 acres (1.2ha) an hour. It was too late; American makers had been producing self-propelled machines since the Massey-Harris 21 of 1940; they had, by 1958, accumulated nearly thirty years of research, field experience and production expertise which made them the indisputable world leaders.

Although the rapidly growing use of combine harvesters from the 1950s had already been the cause of profound changes in farming, further changes were in the offing. The 1960s witnessed the outcome of years of research known as 'the appliance of science'. The work had been undertaken by farmers, biologists, geneticists and agricultural scientists in order to improve the yield of cereal crops. Early results were very encouraging. Within five years the yields being harvested by young, enthusiastic farmers, many trained in the postwar agricultural colleges, who had been co-operating in the research programme of field trials with the new strains of wheat, produced double the number of ears carried on a single stalk. These improved strains, together with the use of dedicated pesticides and efficient fertilizers, quadrupled the annual yields which had been considered good only fifteen years previously.

The much heavier, denser crops and freedom from weeds, following intensive research both in Europe and in the United States, resulted in a reappraisal of the application of the combines that would harvest the crops. The result was intensive competition between combine-makers, each trying to establish a lead in an expanding market. 'Combining' became the normal method of harvesting in Europe, as it had been for at least two decades in the United States.

The development of the combine divides into three clearly defined phases. First was the heroic age of the magnificent forty-horse outfits that were soon to be developed into the far more practical tractor-trailed units, on which much development work was performed in the 1930s to produce the powered take-off tractor-towed reaper-thresher machines. The classic of this period would be the 1938 Massey-Harris 5-ft (1.5-m) Clipper; it was possibly the first reaper-thresher to be known as a combine harvester.

The second phase in the development of the combine is perhaps the definitive one: the introduction of the self-propelled combine. Again, Massey-Harris have a strong claim to classic status as a result of producing their Model No. 21, the machine of the

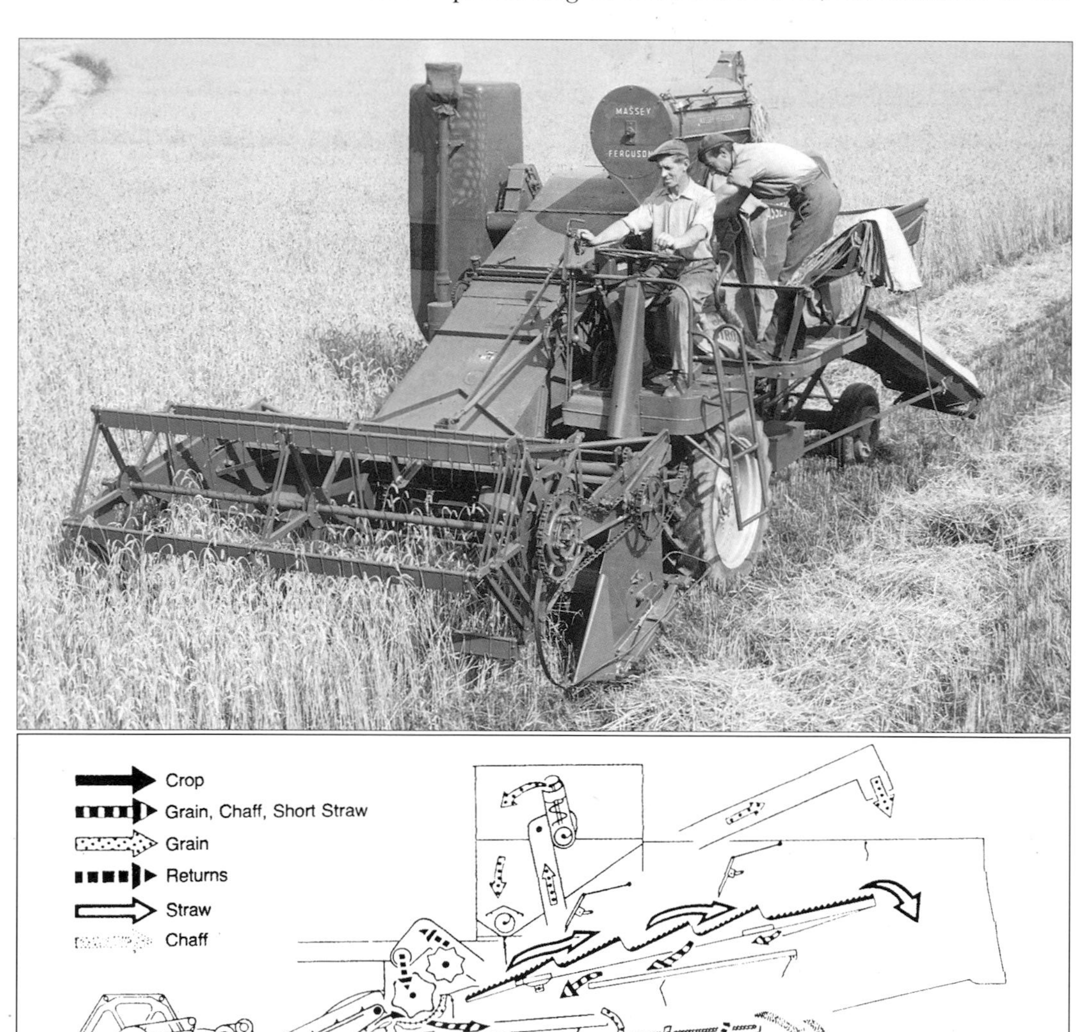

Left A later Massey-Harris combine of the 1950s. The photograph, taken on an English farm, shows that there was, as yet, no attempt to 'style' combines. The driver sits in the open, seemingly surrounded by whirling machinery and exposed to all the dust inevitable during harvesting. The second man is bagging the threshed grain.

Below Although depicting a late combine, this diagram essentially shows the path of the crop through a typical machine. The standing crop is guided to the cutter bar by the reel and, when cut, is lifted by an endless conveyor to the rotating rasps which strip the grain from the straw. The grain is sucked into the grain tank or conveyed by trunking to a truck running alongside. The straw is passed to 'walkers' which shake any remaining grain from the stalk, before dumping them and the chaff from the machine, to be either burned (now illegal in Britain) or later collected from the stubble field.

Because modern combines have to reap the fully ripened crop, good weather is essential. To take advantage of what is often a fleeting weather 'window', farmers are prepared to work round the clock to harvest the grain, as this atmospheric photograph of the harvesting of oats shows.

famous 1944 'Harvest Brigade' enterprise. Tens of thousands of these machines were produced. Those shipped to Britain in 1942 became a benchmark for all the contemporary combines to be produced in the postwar years. Practically all the best of the early machines were made in the United States and, by 1948, 1,500 Massey-Harris 21s were in use on British farms (though Massey-Harris had to compete, by that time, with other excellent combines, such as the McCormick-Deering No. 62). This machine, though not self-propelled, remained in production for ten years from 1941, over 40,000 being sold worldwide. It was one of the last tractor-towed combines to be sold in large numbers; self-propelled machines were the type for the future.

Most of the machines produced from the late 1950s to the 1980s were either clones or developments of the best of the early self-propelled combines, among them some with a serious claim to be classics. There were, of course, improvements and special features announced by optimistic makers along the way, some of importance. The result was that in Britain where, before the war, the number of combine harvesters in use was under a thousand, it was becoming clear that the market in the 1950s would be brisk. To meet the expected demand, American companies like Massey-Harris opened a European factory in Kilmarnock, Scotland in 1949; Allis-Chalmers began, in 1951, to make their All Crop 60 combine in Britain, a rather dated (though, at £650, inexpensive) machine that had first appeared in the United States as far back as 1935. American and British makers together sold machines in numbers sufficient to double the ownership of combines in use on UK farms between 1948 and 1950 to more than ten thousand.

A decade later, the 1960 figure had risen to 50,000, with British farmers alone having no fewer than twelve different makes offering thirty-two different combines, ranging in price from £650 to £4,000, from which to choose. When one considers that the rock-bottom cost of a combine in the 1960s was Allis-Chalmers's All Crop at £650, the figures are the more remarkable. In 1961, the Danish maker, J. F. Farm Machines, offered an inexpensive option to the conventional trailed or self-propelled combine, which was the 'wrap-around'. This combine was laid out in the form of an 'E' with a tractor connected fore and aft by linkages to the central bar of the E. The makers claimed that the tractor could be attached to the combine by one man in five minutes. The tractor provided the propulsion via its wheels and powered the combine's machinery from the power take-off. Cutting widths of 6–10ft (1.8–3m) were possible, but a powerful tractor, of up to 50hp, would be required for the largest cutters.

J. F. Farm Machinery was not the originator of the wrap-around concept. The first combine the Claas Company had made, in about 1935, was a wrap-around. This used a Lanz tractor. Claas was not the last; Ferguson was to offer a wrap-around combine for its very popular TE 20 series of tractors, which were perhaps a little underpowered for the task; in the event the Ferguson wrap-around never enjoyed the sales of the Danish machine, of which some thousands were sold throughout Europe. Although popular with smaller farms, the wrap-around system had faded out by the mid-1970s.

By the mid-1960s new names appeared. The German firm, Claas, which had made its first trailed combine in 1936, produced its self-propelled 'European' combine with a small (by American standards) 7-ft (2-m) cutter. This machine was powered by a 45-hp diesel engine. As the years passed it became noticeable that the power of the engines of combines crept up and up relentlessly. By 1962, a Swedish combine, the impressively named Bolinder-Munktell Viking 6T 1000, had an 80-hp engine and a price tag of £3,000 – a considerable sum at the time. The American maker, John Deere, continued the trend of US companies with plant in Britain by building an Anglicized version of the combines that had been proved successful in the United States, the main difference being the smaller width of the cutters (from around 16ft (4.8m) for the American market down to 7ft (2m) for use on European farms and roads). The prospect of meeting a 16-ft (4.8-m) combine en route when rounding a corner on a narrow country lane was a sobering thought.

The 1964 Lely Fisher-Humphries Victor had an 18-ft (5.4-m) cutter but, in deference to British lanes and farm gateways, the cutter bar and reel was divided into two halves, which were turned upright by hydraulic rams. This reduced the overall width to under 10ft (3m) for travelling on the public highway. The Victor was claimed, at the time, to be the world's largest combine, with an output of up to 17 tons per hour. This combine was one of the first to offer the option of a cab for the driver; it seems strange that the combine had to wait for long for this feature. Archive film taken in the 1960s in the United States shows combines working in clouds of dust. In one such film, the driver looks as if he were about to take off in an aircraft from the First World War: he has a leather helmet, goggles and face mask to protect him from the swirling chaff. Combining on a 1,000-acre (400-ha) field with a following wind must have been, at the very least, a most unpleasant experience, yet it took twenty years for cabs to be fitted generally to combines.

The advances in efficiency, reliability and the availability of combines with such a varied range in capacity and price made combining a viable option for practically all arable farmers with 200 acres (80ha) or more to harvest. Despite all the hardship and worry inevitable with harvesting, even with a good combine, many farmers actually enjoyed it. Here is what one farmer, Jack Dickenson, thought about the annual chore:

> Listen, I've never had a summer holiday. You wait and worry all year about those fields. Just the husky smell of that warm wheat makes me feel happy and satisfied, and the roar of the diesel in the combine starting gets me excited – plenty of time for a holiday after the harvest.

If the power and the detail complexity of the 1960s generation of combines was increasing, so was the cost of buying one. This was a cost that was roughly the equivalent of the annual profit of a sizeable farm, and therefore a very substantial outlay for a machine which was to spend most of its life standing idle and silent in a barn. John Proudfoot, a farmer, put it this way:

> To the farmer the combine is probably his most important investment yet it sits gathering dust most of the time. At harvest time, he will run it for around the three weeks that are the culmination of all his work across the seasons. There is absolutely nothing more satisfying than seeing full trailers of freshly threshed wheat and barley being led off the field to the drier.

Combines might only be active for three weeks or so, but they did require a lot of maintenance. They had a maze of belts, shafts and gears; one leading make had no

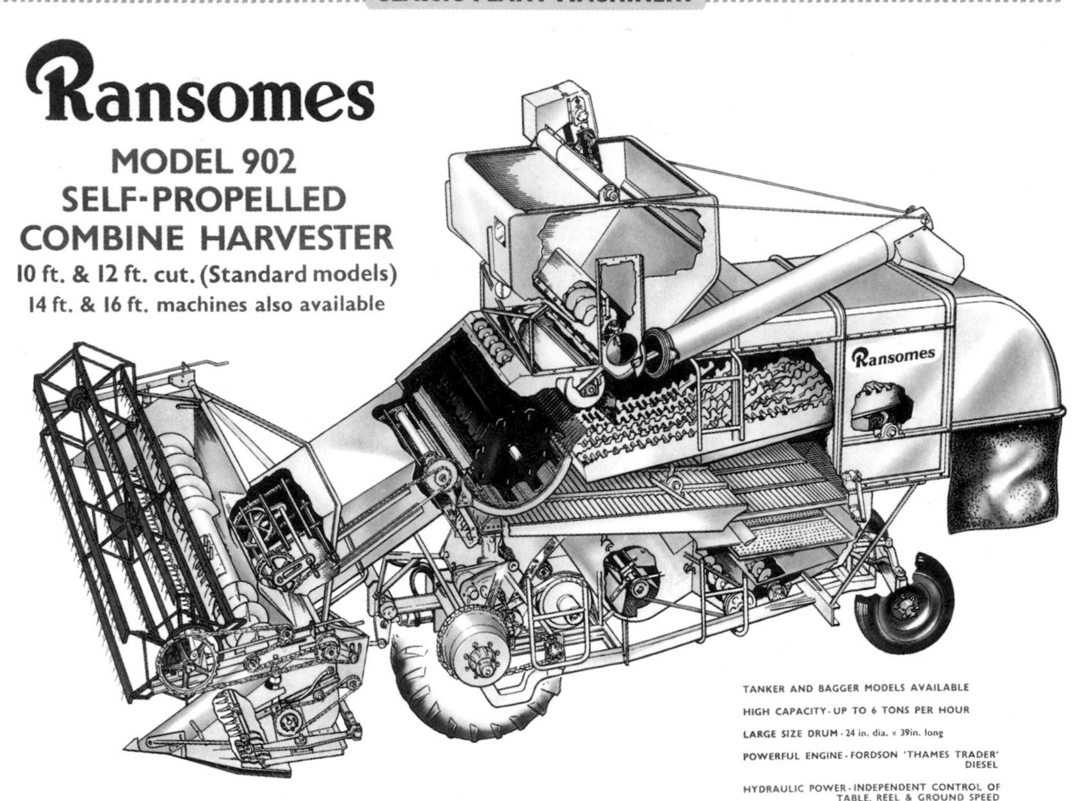

Ransomes, one of the oldest names in British agricultural machinery, entered the combine market as late as 1952. Their Model 902 self-propelled combine of 1970 was the penultimate offering as the company ceased production of combines in 1976. The cut-away drawing shows the rasp bars and straw walkers clearly.

fewer than eighty grease nipples requiring regular attention. By the 1970s combines had power steering, disc brakes and adjustable clearance of the cutters, and were also offered with a self-levelling sieves option and diesel engines of up to 125hp. Work rates of between 6 and 10 tons an hour were commonplace. Most could reap a wide range of crops: wheat, maize, rape, sunflowers, beans, grass and flax. Claas even offered an option of a caterpillar-tracked drive for the harvesting of rice!

By the 1970s combines built and marketed in Britain were no longer cut-down versions of American domestic machines adapted for the smaller farm but designed from the outset specifically for that purpose. The names of the market leaders began to change. Some were old ones, some new, some the result of boardroom mergers, and others were long established. Claas, Case and the Belgian maker, New Holland, began to dominate an increasingly international market. Ransomes, Sims & Jefferies of Ipswich, which was one of the oldest names in agricultural engineering, entered the market with the Swedish-designed MST 42 trailed combine in 1953 and progressed to the self-propelled 902. They made their final combine, the Crusader, and then, sadly, ceased production in 1976.

Most of the combines produced in the 1970s and into the 1980s were variations on the basic theme, using rasp bars, concave sieves and the straw walkers that dated back to the first threshing machines a hundred years previously. The American company International-Harvester produced an axial-flow system in 1979 that had been in use in the United States but was new to British farmers. This required a powerful engine and the one that was fitted topped the 200hp level. Axial-flow combines look like any other externally, as they still use the same cutting system, but the cut wheat is fed by an impeller to a long, rotating cylinder that is not across the harvester, as are the classic

One typical state-of-the-art machine, the Massey-Ferguson Field Star system, justifies the aircraft analogy in that it employs global positioning system (GPS) techniques, developed originally for the US military and used during the Gulf War. The Massey-Ferguson has a powerful on-board computer that constantly monitors the flow of grain as it passes through the elevator into the tanks. The flow-rate information is integrated with the GPS which continuously plots the harvester's position in the field, accurate to about 1 metre (3 ft-3in).

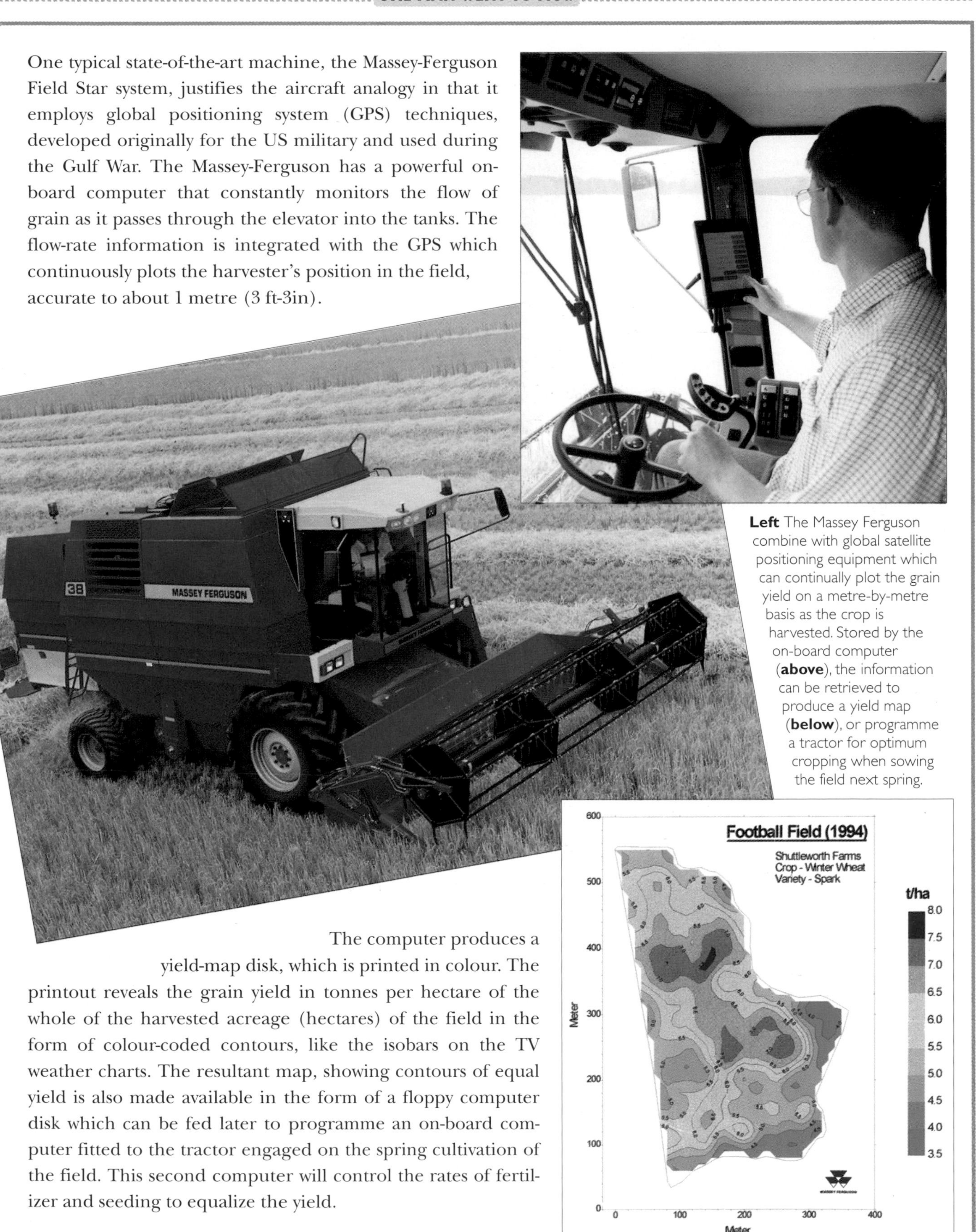

Left The Massey Ferguson combine with global satellite positioning equipment which can continually plot the grain yield on a metre-by-metre basis as the crop is harvested. Stored by the on-board computer (**above**), the information can be retrieved to produce a yield map (**below**), or programme a tractor for optimum cropping when sowing the field next spring.

The computer produces a yield-map disk, which is printed in colour. The printout reveals the grain yield in tonnes per hectare of the whole of the harvested acreage (hectares) of the field in the form of colour-coded contours, like the isobars on the TV weather charts. The resultant map, showing contours of equal yield is also made available in the form of a floppy computer disk which can be fed later to programme an on-board computer fitted to the tractor engaged on the spring cultivation of the field. This second computer will control the rates of fertilizer and seeding to equalize the yield.

rasp bars, but set at 90 degrees to the cutters. This cylinder, typically running at 950rpm, has spiral rasps around the outer circumference, pressing the crop against a cylindrical concave which strips the grain. The straw falls conventionally through sieves as in the rasp-bar system. It is claimed that the axial-flow system has a higher work rate than the conventional methods and causes less damage to the straw. Another variation of the age-old cutter-bar system was an adaptation of a technique popular in Australia: the stripper harvester. Strippers developed in Britain in the 1980s have, in place of the cutter bar, a rotor running anti-clockwise that typically has eight rows of V-shaped teeth, each of which has a keyhole at the apex. The rotor is set to the height of the heads of the crop being harvested. These are stripped by the rotor teeth and passed to a conventional threshing cylinder and concave by an elevator, leaving the straw standing to be dealt with later. In 1988 a British company, Shelbourn Reynolds, offered a 'bolt-on' stripper unit which, it was claimed, could fit most combines and which could harvest a number of crops in addition to wheat. Although grain strippers have remained popular in Australia, the technique made little impact in Europe, one problem being the disposal of the standing straw. The traditional burning of straw and stubble in the fields had become illegal in Britain; the smoke drifting across the new motorways was considered an unacceptable hazard to high-speed traffic.

The late 1980s witnessed what might be called the end of the second classic era. The final form of the mechanical combine was by then in its most developed form. Combines were available in a wide range in size and cost, with small machines capable of 5 tons an hour and medium machines able to harvest up to 7 tons an hour. Most had comfortable (even air-conditioned) cabs, adequate diesel engines, hydraulic brakes, reel height and speed control, power steering and a full set of powerful lights. Harvesting was no longer an exclusively daylight operation. In 1980, the German Claas Company, the only known combine-maker to have had their factory destroyed by enemy action in the Second World War, celebrated the construction of the 300,000th combine made at their works at Hersewinkel.

In the United States, very large combines were in use, often harvesting the prairies with six machines in echelon. Makers like Massey-Ferguson, New Holland and Claas began to offer European versions of what might be called 'super combines', which owed their sophistication to American experience. These large, 300+ hp machines are the combines of the future. They are by far the most advanced to date and set a standard that should last until well into the twenty-first century – they can harvest 10 acres (4ha) an hour.

What will be the final form of the combine? The major advance lies in the control systems of the twenty-first-century combine. Today's driver has become a systems manager; his air-conditioned cab now resembles nothing so much as a jet aircraft flight deck because the function of the whole machine is computer-based and relies on advanced electronics. The Massey-Ferguson 'Fieldstar' system, for example, relies on computer-driven global positioning, using satelites which can plot the yield of a crop as it is harvested to within a metre, and later will adjust the fertilizer/seed required next spring to obtain a uniform yield. In practice it takes more than one season to achieve a uniform yield, assuming that the field is not concealing a lost Roman villa, or a crashed aircraft from the Second World War, which would permanently inhibit crop growth. The whole operation of the later generation of combines is done by electronic screens displaying all the relevant information the 'driver' requires: flow rates, reel and cutter heights and forward speed can be constantly monitored as they automatically set themselves to optimum. All the controls are push-button in air-conditioned, sound-deadened cabs, providing a working environment that the helmeted and goggled pioneer combine drivers would find unbelievable. There is a downside. These machines are seriously expensive; at the top of the range they may be nearly £250,000. Obviously that level of investment is not one that a man with 200 acres (80ha) can be interested in. It has resulted in big land and estate owners grubbing up hedges to provide mini-prairies for these big machines to harvest. Ecologists are alarmed, many would say rightly so, but farming is an industry like any other and the fields are the factory; farmers are not national park-keepers. The debate continues.

What of the 200-acre (80-ha) farmer? There is a growing conviction among smaller farmers that contemporary combines are too expensive and too clever, and that there is, therefore, a need for an alternative. This would be an all-mechanical, simple, affordable machine, possibly an updated version of the wrap-around combines of the 1970s, which could take advantage of the powerful tractors that are today standard on most farms.

Complexity and cost aside, the efficiency of the latest generation of combine harvesters is impressive. Grain loss, which in the days of reaping and threshing (gleaners not withstanding) was considered acceptable at 40 per cent today averages 4 per cent, a figure that is probably irreducible. There is one final thought. Despite all the advances, improvements, electronics, computers, GPS and so on, if the ghost of Patrick Bell, who exhibited his mechanical 'reaper' at Crystal Palace in 1851, could see the latest combine, the reel and cutter bars, at least, would still in principle be the same as his original design.

This Claas combine c.1950 is in sharp contrast to the latest 'state-of-the-art' 1997 machines with satellite and electronic control, yet the early machines did work well. Many farmers with smaller holdings think that a return to these simpler (cheaper) machines without 'whistles and bells' would suit them well, even in today's computer-controlled industry.

6

CHAPTER FOUR

Hands off: The Era of Mechanical Handling

In 1948, Britain was only just showing signs of recovery after the six long years of the Second World War. There were still shortages of practically everything; residual rationing was still in force and austerity was the watchword as ex-servicemen and women began to pick up the threads of civilian life. Industry, too, struggled to return from war production to fulfilling civilian needs. Trade exhibitions were held and many firms, eager to promote their wares in the postwar markets, participated. One such exhibition, held at London's Olympia, was unusual in that it had a title that was unfamiliar to most people: 'The 1948 Mechanical Handling Exhibition'. What, people asked, was mechanical handling?

Unlike other classic plant featured in this book, mechanical handling was, in 1948, a recent development that owed little to prewar practice. It is true that in the major 'smokestack' industries there was always a measure of mechanical handling, simply because the weight and size of components used in shipyards, foundries or locomotive works could not be effectively manhandled by those universal, age-old engines of human and animal muscle power. A forest of cranes surrounded a ship on the slipways; travelling cranes lifted and carried ingots or locomotives from place to place in steel mills and railway erection shops. Such large-scale mechanical handling aids might be correctly described as the process and consequence of heavy industry rather than mechanical handling in the present context, which is the substitution of mechanical muscle for human muscle. This involves machines that can lift a range of loads formerly handled by men, who either worked singly or in groups without any mechanical aids other than a wheelbarrow, a hand-drawn truck or simply a ladder and two hands. Or it might involve collecting stores from a warehouse, moving parts in a light-engineering works, loading lorries or carrying parcels and passengers' luggage at railway stations.

The 1948 Olympia Mechanical Handling Exhibition was mounted to promote just such mechanical aids. The idea owed much to the American services in Europe, the US Air Force in particular who, during the war, had used small fork-lift trucks to shift spare aircraft parts around their bases. One British firm showing at Olympia was Lansing Bagnall. The company was exhibiting the prototype of its Model P, a newly designed pedestrian pallet truck. It was electrically powered and the company set great store by it; as the exhibition opened its doors the truck was only just ready. Indeed, the designer

Modern fork-lifts can utilize warehouse space in a way which was unthinkable for their pioneers.

Mechanical handling at London docks in August 1940. Though still labour-intensive, without the aid of cranes the unloading of ships would have been a very protracted affair. In wartime, when this picture was taken, speed was vital in turning round a ship. In peacetime, too, it was essential as ships tied in port were not earning revenue.

and staff from the company had been up all night assembling the truck at their small factory at Isleworth and it had only arrived at the exhibition as the doors opened. The truck caused great interest, not least because when first demonstrated for a potential customer, the truck proceeded to demolish the flimsy plywood stand as the luckless driver struggled to control it. In the haste to get the machine to the show, a very tired mechanic had cross-wired the forward-reverse switch, causing the truck to set off in the opposite direction to that intended. The fault was quickly rectified and the Lansing Bagnall Model P, the first electrically powered pallet truck to be designed and built in a British factory, was launched. Although it only had a pallet lift height of 3in (7.5cm), it was to prove a great success both at the 1948 Olympia show and subsequently in production, thousands being sold during a seventeen-year production life. The Lansing Bagnall Model P could lay claim to be a pioneer in European mechanical handling.

The company had only been formed during the war in 1943 with seven people constituting the entire workforce. However, the origins of the firm go back much further, to 1920, at a time when mechanical handling was almost unknown in Europe. In the United States, on the other hand, it was well established, another possible consequence of the shortage of low-waged, unskilled labour that was to speed the development of mechanization. This had also been the case with the introduction of American steam-powered excavators, made necessary in the first instance through a chronic shortage of low-waged manpower required to build the country's railroads in the nineteenth century.

In Britain, in 1920, F. E. Bagnall, an engineer who had been to the United States, had seen small electrical battery-operated platform trucks performing as 'mechanized muscle', moving heavy loads about factories and warehouses with greater efficiency and speed than hand-drawn trucks could possibly manage. The platform trucks that so interested Bagnall were manufactured by the Lansing Company of Michigan. True to the tradition of so many companies in the United States, Lansing was the name of the town in which the company had its plant. Bagnall lost no time in securing the UK agency for the Lansing electric platform trucks and a primitive, very heavy, battery-operated tractor that had a low-mounted, fixed fork on the front to carry drums or similar loads; this was the Lansing Model D of 1925. The imported trucks sold well to railway companies who used the forward-control Model O platform truck extensively on stations and in goods yards; paper mills and other plant with a high risk of fire also bought the Lansing electric models. The restriction imposed by the limited capacity of the batteries was not severe because the trucks worked in a limited area with overnight recharging. There was no attempt to style the Lansing trucks; they had solid tyres with the drivers standing in the open at the front, in the case of the Model O platform truck, and at the rear of the Model D fork tractor.

In 1930 the American Lansing Company decided to cease the manufacturing of its, by American standards, rather dated platform trucks and tractors. Sensing that the British market was still viable, F. E. Bagnall acquired the British manufacturing rights to the Lansing designs, to be constructed under contract by an engineering company at Stanford le Hope, Essex. This was Thomas Allen Ltd; the products were to be sold under the Lansing name.

The London Midland & Scottish Railway (LMS), the largest of the four main British railway companies, which had been using the Lansing electric platform trucks in some numbers, asked Bagnall to design for them a small petrol-powered tractor to be used to

The Lansing Bagnall stand at the 1948 Mechanical Handling Exhibition, the first to be held in Britain. 'Electric Traction in a New Form' the placard on the stand proclaimed. It proved to be true: the display of these early mechanical-handling trucks created a sensation at the Olympia exhibition.

tow hand-operated platform trucks. These were a common sight at all railway stations in Britain for luggage, mail, newspapers and the light freight and packages that were to be loaded into the luggage vans of parcel and passenger trains.

The tractor built for the LMS, called the Imp, appeared in 1934 and was soon working at main-line stations throughout that railway's system. It was an advance on the earlier, pioneering Lansing platform trucks. The Imp had neat, small wheels shod with fat pneumatic tyres and, in true tractor tradition, the rear wheels were about a third larger in diameter than the front ones. The layout was also classic; the driver sat over the rear wheels with the neatly cowled petrol engine in front. Effectively silenced, the Imp caused no offence to rail travellers and could, and did, tow a surprisingly large number of standard railway platform trucks in train around the major LMS stations, Euston being one of the first. The success of the Imp encouraged Bagnall to form, in 1937, a new company: Lansing Bagnall & Co. Ltd, with F. E. Bagnall himself as managing director. The Lansing Bagnall Company decided that all future production of the Imp and the platform trucks would be in-house at their leased factory at Isleworth in Middlesex.

In the first year, twenty-five Imp tractors were built and sold. Several of the electric platform trucks were also built, but it was far from the best of times with war and rumours

Soon to become a familiar sight on the mainline railway stations of the LMS: the first Lansing 'Imp' tractor, seen in 1934 towing a train of baggage trolleys along platform 1 at Euston station.

The Lister factory in rural Dursley. The steep, badly surfaced roadways between the company's railway sidings, which drained the batteries of the electrical platform trucks, caused the company to evaluate a small petrol-driven 'Auto Truck' in the mid 1920s, when this picture was taken. They bought not only the truck but the company that made it. The result was the world-famous 'Lister Auto Truck'.

of war dominating the newspapers of Europe. In 1938 the Munich crisis took place with the British Prime Minister Neville Chamberlain flying to Germany to appease Hitler. The markets were understandably nervous; it was a time of great uncertainty with trade slumped generally, and orders for the Lansing Bagnall trucks and tractors dropped away. By September 1939 Britain was at war with Germany; it was a very difficult time for small companies not directly involved in war production or essential work. In 1940 F. E. Bagnall resigned; his successor died the following year. The small company struggled on under wartime shortages and regulations to become even smaller, relying on the supply of spares for the Imp and platform trucks to keep solvent. By January 1943 it had shrunk to a payroll of just seven people and faced the prospect of imminent bankruptcy. The owners of the leased Isleworth factory, no doubt keen to find a more promising tenant, suggested to another engineering client, J. E. Shay Ltd, nearby at Mortlake, that in view of the parlous state of Lansing Bagnall it might be acquired if a modest offer was made.

J. E. Shay was a small firm of precision engineers owned by two partners, Emmanuel Kaye and J. R. Sharp (the name 'Shay' was made from the two partners' names) who, although engaged on wartime engineering contract work, were looking ahead to the postwar period and some engineering-based product that they might manufacture. In short, on 16 April 1943, J. E. Shay Ltd became Lansing Bagnall (1943) Ltd, now owned by the two consulting engineers. However, the two directors must have wondered if they had made the right move – would there be a market for mechanical handling after the war, and what level of competition would they face?

The answer to this was: very little. The existing competition came from the very old-established, respected company of R. A. Lister of Dursley. Lister was, and remains, a name associated with reliable diesel engines and diesel-powered electrical generator sets offered over a wide range of outputs. In the mid-1920s, following another piecemeal enlargement of the Lister factory, it was clear that, despite the extensions, the use of horse-drawn wagons to transport heavy engine castings and parts around the factory was causing difficulties due to restrictions of space. It was decided that it was time to dispense with horses and investigate an alternative, twentieth-century mechanical method of handling the castings and parts within the factory. Electrical platform trucks were evaluated, almost certainly imported Lansings. These were not deemed satisfactory, the heavy loads and a steep hill within the factory causing the batteries to require frequent recharging. Another company, the Auto Mower Engineering Co. Ltd, was approached and willingly loaned to Lister, for appraisal, one of their small petrol-engined 'Auto Trucks', designed by a George Grist of Norton St Phillip, a Somerset village, whose main claim to fame was its local pub, the George Inn, dating from the thirteenth century. To what extent, if any, the George Inn influenced the Auto Truck design is unknown, but its performance was certainly to influence the Lister management. The truck when tested could, without difficulty, carry a 1-ton casting not only on a relatively smooth and level factory floor but also up the steep and poorly surfaced road between the foundry and the company's railway sidings. Not only that, but its clever design allowed for great manoeuvrability, an important feature when working in confined spaces. The testing over, Lister, instead of buying the Auto Truck, in fact bought the Auto Truck Engineering Co. – patents, name, designs and all manufacturing rights. Though possibly not at that time aware of it, Lister had, about 1925, entered the nascent mechanical handling business.

After some detailed revision of the original design, the first Lister Auto Truck was sold the following year. From the outset the Lister Company decided that in order to keep down the price of the truck to a very reasonable £100 it would only be produced as a standard model. The design was exceptionally good; essentially, it was a three-wheel chassis with a single driven wheel at the front. This wheel was fixed; above it was the single-cylinder air-cooled engine within a tall cowling, which also contained the fuel tank, the simple controls and the steering. The drive to the engine was via a clutch and a single forward gear. The clever feature of the design was that the tall engine-transmission structure on its single wheel was, at its base, mounted on the chassis with a wide, 360-degree bearing, rather like the turret of a warship's gun. This simple arrangement allowed the entire motive section of the truck to be rotated through 360 degrees, offering exceptional handling in confined spaces. It could turn in its own length and even, if required, reverse as well. The driver sat behind the rotatable engine mounting and steered it and the truck with a wide handlebar attached to the movable structure. Behind him – or her (during the war, women often drove the Lister) a flat, low platform offered easy loading and unloading.

The Lister Auto Truck was an instant success. Orders poured in; by 1938 10,000 had been sold at home and abroad. Despite the firm's avowed intention only to produce a single standard model, modification soon followed. The fixed gear gave way to a gearbox, the clutch was improved, tyres enlarged and the rotating mounting was redesigned with heavy roller bearings replacing the original plain bearings. At the British Empire Trade Exhibition in Buenos Aires, held in 1930, a number of Listers were converted to carry passengers around the exhibition; these trucks also later served other exhibitions. Other modifications followed: longer or shorter load platforms, and load platforms higher or lower than standard. One keen customer had a Lister Auto Truck converted to a narrow-gauge railway locomotive; it proved to have a tractive effort measured at 24 tons, from a machine that weighed 1½ tons itself. Several hundred standard Lister Auto Trucks were ordered by the Great Western Railway; substantial orders also came from the War Department and the Admiralty. During the war, thousands of the Lister trucks worked on mechanical handling in ordnance depots, service workshops and in the many munitions and aircraft factories all over Britain, many being modified designs tailored to the precise needs of individual customers. In all, over 3,000 different versions of the truck had been made when, in 1973, production of the Lister Auto Truck was transferred to Compton Electricars at Tredegar, Monmouthshire, South Wales.

Despite the undoubted success of the Lister Auto Truck, it did not, during its forty-seven years in production, change its basic function – that of a handy, adequately powered, flat-bed platform truck. It was, after all, in the nature of a sideline for the company. The Lister Company was, at core, a diesel-engine manufacturer; the mechanical handling business came as a consequence of the firm originally requiring a suitable truck for use in their own factory. The Auto Truck they tested in 1925 was so well suited to the work and so lacking in competition in the market that, through business

By the early 1930s the Lister Auto Truck was established as a major unit of mechanical handling in and around factories. On the occasion depicted, the Duke of York, soon to become King George VI, drives an early example, with bowler-hatted directors in attendance. The gentleman on the right seems in some danger of having his left foot run over by the royal driver.

acumen, the Lister board saw the commercial possibilities which seem to have eluded the original Auto Truck company management. Though developed by Lister over the years, the truck was to remain the sole, though significant, contribution made by the company in the field of mechanical handling.

The company that was to dominate mechanical handling in the postwar years was Lansing Bagnall. On acquiring the nearly defunct company in 1943, the two partners, Emmanuel Kaye and J. R. Sharp, must have considered themselves extremely unfortunate because, after their tiny workforce had produced an encouraging fourteen trucks in the first six months at Isleworth, came disaster. The all-powerful wartime Ministry of Supply wrote to inform the two directors that all production must cease forthwith and that, as a consequence, no further supplies of steel or other proscribed materials would be forthcoming. It is hard now, fifty years later, to understand the absolute power over industry that the ministries wielded during the years of the Second World War.

The Lansing Bagnall board appealed against the, as they saw it, arbitrary order. Happily, the appeal was upheld and the company could resume production. However, the processes of democracy take time and the company lost eighteen months of production; it was not able to resume construction of the trucks and tractors until December 1944. The firm had kept solvent during the enforced interregnum by transferring machine tools from the parent company of Shay Ltd and undertaking a share of the precision engineering work. Paradoxically, one ministry having effectively closed down the Lansing Bagnall truck business, another, the Ministry of Aircraft Production, unwittingly threw them a lifeline. The contract work that the Ministry of Aircraft Production had offered to Shay Ltd involved the production of precision parts used in certain control systems of RAF bombers. The work was both urgent and extensive, and was sufficient to keep both the small factories at full stretch.

Although the unforeseen war work was the salvation of Lansing Bagnall, it would clearly cease once the war was over. By late 1944, with the Allied armies inexorably advancing into Germany, the question now was not how the war would end but when. The two partners of Lansing Bagnall were therefore prudently considering the options open to them in the postwar mechanical handling market. Several propositions were considered with hopeful inventors interviewed. One man they talked to was A. R. Arnot, an electrical engineer, who impressed them with his original ideas and was subsequently invited to join the company as designer. At Isleworth, the remaining months of the war were largely to be devoted to the future, and the design and construction of a prototype new mini-towing tractor, the Lansing Bagnall Model A that would replace the prewar Imp, now dated. Late in 1944, the LMS railway approached the company with a contract to recondition the existing, elderly prewar Imp fleet; that work neatly tided the company over from war to peacetime work until mid-1945, when the new Model A would be introduced. The old, 1925, Lansing battery-powered platform truck design had also been revised, updated and catalogued as the Lansing Bagnall Model O. Twelve were ordered 'off the drawing board' by the Belfast Steamship Company to supplement six prewar Lansing platform trucks (three of the six were still operating in 1968 after forty-three years of service).

The Model A was the first tractor that was to be designed from the outset by the new company; it owed little to the earlier Imp machine. To say it was designed at Isleworth is true only in the sense that it was not designed elsewhere; at Isleworth there was no drawing office, and no draughtsmen at work on their boards. The general arrangement of the new tractor and trucks was simply drawn in chalk on the factory floor as in a shipyard moulding loft. The 'design' was very largely empirical, Arnot and the two partners erasing and redrawing until the full-sized chalked outline on the floor 'looked right'. A graphic artist was then hired to turn the outline into a three-dimensional impression of the tractor. The same artist also drew the winged Pegasus that became the company's now famous logo.

The company celebrated VE Day, the end of the war in Europe, by putting the finishing touches to the first Lansing Bagnall Model A tractor that was to be announced in June 1945. It was demonstrated in October that year to a hard-bitten collection of officials from the procurement office of the Ministry of Supply, which two years earlier had ordered the company to close down. The men from the ministry were accompanied by representatives from the Royal Army Ordnance Corps who were keen to supplement their existing Lister Auto Truck tractors with another type which had more pulling power. The Lansing Bagnall Model A proved the point by towing a load of 23 tons, which consisted of an army tank transporter with a Dodge 3-ton truck on board. Fifty Model A tractors were ordered on the spot for use in Ordnance depots. Lansing Bagnall, the total staff now numbering twenty-two, became the market leader overnight in the postwar British mechanical handling business.

The Lansing Bagnall Model A tractor, introduced in 1945, was evaluated both by its many potential users (including the Army Ordnance Corps) and by railway companies while on a countrywide tour. Here, a model A is tested at the Crewe works of the LMS railway, hauling five loaded platform trucks without difficulty.

In the first production run of the Model A it was powered by a four-cylinder Hotchkiss petrol engine, a later version of the engine that had powered the immortal 'bull-nose' Morris Cowley. The new Lansing Bagnall tractor was extensively demonstrated to industry, towed to the sites on a low-loader trailer behind a prewar Humber car. Orders came and a hundred Model A tractors were built at Isleworth during the first year of production. One order was unexpected; it was from a golf club which specified extra large tyres because the tractor was bought to tow the club's grass mowers. Other golf clubs soon saw the results and more orders followed for the Lansing Bagnall 'mower'. During the first year's production the tractors were exported to countries as far apart as New Zealand, South Africa and Sweden; the first international orders for a design that was to remain in continuous production for the next twenty years until the Model A was finally withdrawn in 1965. By that time several proprietary engines had been fitted, including a Perkins diesel engine.

That first year had been encouraging. However, selling a hundred tractors was not really doing more than scratching the surface of mechanical handling when compared with the world leaders in that field: the Americans. Aware of the situation, the two partners, Kaye and Sharp, left the Isleworth works in the hands of the production manager and set off for the United States and a tour of the best in mechanical handling. What they saw impressed them and confirmed what had been seen of the fork-lift trucks used on their bases during the war by the US Air Force in England. After the tour the partners came back to Isleworth with an agreement to act as the UK concessionaires for the electric fork-lift trucks made in the United States by the Baker Rauling Company of Cleveland, Ohio. The first Baker fork-lift trucks arrived by sea in January 1947.

The concession was useful but possibly unnecessary for, by the start of 1947, A. R. Arnot had designed and patented a very efficient electrical control circuit which incorporated a shunt-wound motor. This offered several advantages: liner acceleration and regenerative braking, whereby the motor, when required, can be switched to act as a generator loaded by a resistance, which greatly assists in braking the truck. Regenerative braking had been used for many years on trams but Arnot's system was the first to be applied to battery-powered mechanical handling trucks. Petrol-powered platform trucks in the 1-ton class were offered the 'DP', which was an electric version, while the petrol-engined 'DE' incorporated Arnot's patents and had the driver at the rear with the platform in front under his observation. Both the trucks had ample, sectioned pneumatic tyres – unlike the prewar Lansing trucks, which were shod with unyielding solid tyres.

It had been clear as early as 1947, from the interest and the volume of orders the young company was receiving, that the leased factory at Isleworth was too small if the company was to expand. In the immediate postwar years Britain had not recovered from the strain and shortages of the war; steel was controlled and permits were necessary for practically everything. Nevertheless, after a search, a green field was secured as a site outside Basingstoke, Hampshire. It was truly a green field; this is confirmed by a photograph of the entire management staff of Lansing Bagnall (ten) taken on the site in front of two haystacks!

Permission to build was granted in October 1947 and, true to form, the Ministry of Supply promptly cancelled the licence. Appeals followed, and finally work started in the

Although Lansing Bagnall were to become the market leaders in Britain, they began by importing the American Baker fork-lift as is shown in this 1948 picture of a Baker at work in the stores of the Kodak company in west London alongside a Lansing Bagnall platform truck.

summer of 1948. The factory had to incorporate an old cowshed and two ex-army huts. Work started on the newly designed first Lansing Bagnall pallet truck, the first to be built in this country; this was the Model P that was to create a sensation at the Mechanical Handling Exhibition at Olympia. ICI bought four; other orders followed and, by the end of 1948, fifty-one had been sold and delivered. That the pallet truck was an advance was demonstrated by the decline in orders for the improved platform range. Mechanical handling was gradually catching on, even among conservative managers of British factories. The 'P' range was developed to the 'PF' type, introduced in 1950, which had a hydraulically powered lifting fork capable of stacking stores, etc., up to a height of 10ft 10in (3.25m). This was truly an advance. Since the Industrial Revolution, stores and warehouses had been used in a very inefficient way. The heavy items, which had to be manhandled by a number of men, stayed on the warehouse floor. Items that one or two men could reach up to were stacked on racks not higher than about 6ft (1.8m). Above that, there were only items that could be safely handled by a man on a ladder. Since the smaller, light stores or parts were usually the ones most in demand, the constant running up and down ladders was very wasteful of time and no doubt caused many accidents.

By the mid 1950s mechanical handling had become commonplace in every factory in Britain. Fork-lift trucks altered for ever the layout and the efficiency of warehouse management.

STORE IN A COOL PLACE
LIFEGUARD
Andrex

By 1952 it was obvious that the future of the company lay in the provision of fork-lift trucks. The earlier platform trucks that had given the company its initial success were dropped and all future effort was directed to enlarging the range of fork-lift trucks with full hydraulic operation, the rams and jacks of which were to be produced at Basingstoke. The factory was continuously enlarged and production increased; by 1955 the annual vehicle output was increased by 40 per cent, equivalent to no fewer than 1,444 units.

In 1961 a second very much larger factory was built, opened and was soon employing 3,500 people – a dramatic increase over the original workforce of seven with which the partners started in 1943. The policy of the Lansing Bagnall board had always been to develop and expand the range of vehicles offered. The 'PF', with a hydraulically actuated mast, was first shown at the 1950 Mechanical Handling Exhibition and the slogan 'Mechanised Muscle' was coined by the company. The close connection with the railways continued. A 'Transloader' was put into production in 1954; this was a small, pedestrian, battery-powered pallet truck designed specifically to load pallets into railway goods wagons; it was demonstrated at Farringdon Street goods station and displayed extreme agility, especially when working in very confined areas. It found a ready market at home and abroad. Perhaps the most telling of the new types introduced in the late 1950s was the 'Reach Truck', which was the answer to all warehouse managers' problems: this was the Lansing Bagnall Model FRER 2 of 1961, a 2-ton fully developed version of an original fork-lift. It was capable of working in a warehouse aisle only 7ft 6in (2.25m) wide and was fitted with a triple extension mast enabling it to stack up to a height of 23ft (6.9m). A version for cold stores, capable of working at a temperature of -20°F (-29°C), fifty-two degrees of frost, was also offered. Mechanical handling had come of age.

Lansing Bagnall had for some time produced all the electric traction motors in-house at Basingstoke. In 1961 the company had developed a new compact motor placed within the wheels, driving them through an epicyclic gearbox. So successful were the motors that it was decided in 1965 to drop all diesel-powered trucks and concentrate solely on electrical traction. Before considering the current range of trucks mention must be made of the film

Above Although mechanical handling was well established by 1950, the Lansing Bagnall company still had to demonstrate their trucks to often sceptical managers. The managers were probably impressed; however, the body-language of the workers in the picture is not encouraging.

Opposite By the 1960s Lansing Bagnall could offer the first triple-mast reach truck with a maximum reach of no less than 23ft. The 1961 photograph shows an early one at work in a London grocery warehouse, safely retrieving a selected load at near-maximum elevation.

which brought mechanical handling, or rather mechanical mishandling, to a wider public. The film, made in 1958, was *I'm All Right Jack*, starring Peter Sellers, Ian Carmichael, Terry Thomas and eight Lansing Bagnall fork-lift trucks. The film, a black comedy, was a box-office success at the time of its release and, more recently, was very popular when shown on television. The action takes place in a strike-torn factory and the fork-lift trucks figure in many scenes. It was to prove excellent free publicity for the company, and fork-lift trucks were provided by the firm to be placed in the foyers of cinemas showing the film.

In some of the scenes of *I'm All Right Jack*, Ian Carmichael drove his fork-lift truck with some *élan*. In reality, the model used then was only capable of about 6mph (9.6km/h), a speed found to be adequate and safe for most industrial applications. Speed, of lack of it, had never figured to any extent in fork-lift truck specifications. However, in response to an enquiry from Swedish Railways, who wished to replace a large fleet of petrol-engined fork-lift trucks with an electric version, Lansing Bagnall were invited to tender a suitable vehicle. High speed was to be the essence of a contract; the replacement trucks had to match the top speed of the existing petrol-engined units. They had a maximum speed of 15mph (24km/h). This speed is not very difficult to achieve with a petrol or diesel engine, but for a battery-operated vehicle it is very quick indeed; standard battery-powered fork-lift trucks are, for their size, heavy. They make a virtue of necessity by utilizing battery weight as an aid to stability, essential to a high-reaching fork-lift truck.

The high speed required by the Swedish contract entailed the provision of a greatly increased battery weight, requiring an enlarged motor with the inevitable higher current demands. These had a knock-on effect on controls, battery capacity, heavy-duty electrical contactors and control units. In the event, the prototype Swedish 'Revolutionary Rapide' was produced at Basingstoke in just eight weeks. The problems were resolved first by the battery-makers, Exide, who produced a new, very compact battery pack offering the comparatively high voltage of 48 volts, which reduced the current demands. That, together with an efficient new traction motor and an elegant, newly designed control system, enabled the designers to produce an electric fork-lift truck with a speed of 15mph (24km/h) which, in field trials in Sweden, matched their petrol-engined versions in terms of workload. It was also silent and there was total immunity from any risk of fire or carbon-monoxide poisoning when working in enclosed warehouses. The Swedes placed a substantial order and the Revolutionary Rapide also entered general production.

By the end of the 1960s, Lansing Bagnall, soon to drop the 'Bagnall' and become once again simply 'Lansing', were the undoubted market leaders. The range was continuously expanded and developed to meet the customer's exacting needs. In 1989 the company amalgamated with the German Linde Company, and Lansing-Linde, in the third Basingstoke factory, built at a cost of £55 million, still served the needs of a wide swathe of industry in 'materials handling'. In addition to what might be termed the classic fork-lift trucks, the company also offers a small tractor and an airport baggage-handling truck; they have a factory in China and an office in Japan. Lansing-Linde fork-lift trucks work in freezer rooms, but now have an insulated and heated cab for the driver. Computerized Lansing-Linde 'picker' fork-lift trucks that work in supermarket warehouses are programmed to load only selected quantities of produce in a given sequence. The publisher, Penguin Books, has programmed a Lansing fork-lift truck to select packs of a particular title. Trucks are available with four-wheel steering to give maximum manoeuvrability when working, as

Opposite A change of name and a new machine: the Lansing-Linde V10M light 'order picker'. These advanced fork-lifts can be programmed to 'pick' only the required order in a supermarket supply depot, be it a case of beans or mushy peas. Publishers' warehouses, too, use 'pickers' to select a batch of a specific edition only.

37
EPSON
Stylus
Stylus Pro

The Coventry Climax Company, though primarily engine builders, did offer rather large petrol-engined fork-lifts in the 1950s. Here, one displays its power by lifting a Lancia Aprilia saloon. (That late-1930s car would be far too collectable today to be put at risk for a mere publicity photograph.)

fork-lifts often have to, in restricted spaces. The range is from the small 'pickers' up to 52-ton monsters. They are everywhere. As Bob Helbert of Lansing-Linde says: 'Everything you touch, eat, wear has been touched by a fork-lift at some stage – everything.'

There have been competitors along the way. In 1948 the engine manufacturers, Coventry Climax, offered a range of rugged fork-lift trucks that could lift a 2-ton load. They have the distinction of being immortalized by a Meccano Dinky Toy model of their truck, listed as No. 14 of 1949. Coventry Climax trucks were the first petrol-powered fork-lifts produced in this country but, unlike Lansing, which pioneered and developed battery-powered electrical fork-lift trucks to a very high degree, with their primary market seen as working in restricted space in enclosed warehouses, the Coventry firm, producing larger trucks for open sites, only ever offered a basic petrol-engined fork-lift. This truck, though remaining in production for some years, was not to be developed as fully as the Lansing range. This was possibly due to Coventry Climax having, as their major interest, engine manufacturing. Indeed, Coventry Climax engines powered Lotus and Cooper F1 cars in the 1960s, though the maximum speed of their fork-lifts was governed by a much more modest, industrial speed range.

There is another, new and more serious competitor in the form of the Uttoxeter company, JCB, international leaders for many years in the field of excavators and backhoe loaders. They have recently, as they themselves have put it, ' . . . taken a fresh look at fork-lifts'. What they have done is to adopt the telescopic boom, developed for their range of telescopic handlers in use on building sites, to function on a new, compact fork-lift, the JCB Teletruk range. This advanced design, announced in 1997, does indeed look different from all that has gone before. The fork-lift is carried, not on the traditional extending mast, but on a telescopic arm as on modern mobile cranes. The arm is mounted on the right-hand side of the truck, with the fork-lift mounted on the left of the arm. This arrangement allows the operator/driver to have a clear view of the pallet load, unobscured by the usual mast structure.

Another feature claimed by the makers is the ability of the new JCB fork-lift to extend the reach of the arm to, so to speak, load or unload from the 'back of the shelf'. The telescopic nature of the fork-lift arm also gives the operator a very fine control of pallet loading because, with his truck remaining static, the forward placing of the fork is controlled by the hydraulic movement of the arm. Because the telescopic arm, as it rises, describes an arc, the fork is made self-levelling. Rear-wheel steering offers good manoeuvring; the transmission is described by the makers as 'intelligent' and can be matched to individual site and operational requirements. To achieve this, various options are offered: hydrostatic drive and torque converter responses, which can be selected and memorized by an on-board processor. The Teletruk is offered in two lift capacities: TLT 20 at 2,000kg (4,400lb) or TLT 25 at 2,500kg (5,500lb), powered by two engine options, either diesel or low-pressure gas (LPG). The fork-lift height is standard at 4,100mm (123in). The driver sits in a sturdy cab with state-of-the-art instruments and controls that have information and engine warning lights, an hourmeter, clock and, most importantly, a load-moment indicator placed at eye level giving visual warning of longitudinal stability through reach and lift. The visual indication is reinforced with an audible warning. A wide range of attachments in place of the standard fork-lift is available, including an effective bale clamp, shovel and other items. All are rotatable. The compact size of the Teletruk enables it to enter to load or unload a standard container. The Teletruk is claimed to be the only counterbalanced lift truck capable of safely reaching forward below ground level. Mechanical handling has come a long way from the simple battery, pedestrian-operated pallet trucks of the 1940s.

Finally, here is a summing up from the surviving partner of the 1943 company that brought mechanical handling to Britain, the octogenarian and pioneer, Sir Emmanuel Kaye:

> Engineering history is filled with melancholy examples of a lost pioneering spirit. However excellent our trucks were at any stage, it is important that engineers did not believe that any model could not be improved. Technical progress is best achieved with a form of artist's divine discontent.

The highly competitive JCB company recently decided to '... take a fresh look at fork-lifts'. The result is the 'Teletruk': a telescopic-lift truck which can reach into a shelf as well as lifting up and down. The side-mounted lifting arm offers the driver a view unobstructed by the conventional fork-lift masts.

CHAPTER FIVE

Digging Deep: The Excavators

ANY ARCHAEOLOGIST will tell you that there is nothing as permanent as a hole. You can fill it in, tamp it down, plough over it or plant crops in it, but it will remain detectable, probably for ever. Post-holes for Stone Age dwellings can be defined and excavated centuries after all trace of the structures they were intended for have vanished. We have, therefore, the evidence of human burrowing from the dawn of time.

People have always had, it would seem, an irresistible need to dig up the earth. Holes are made for foundations, to bury the dead, to store grain, for wells, for mining for minerals, as culverts for drainage, as trenches, ditches, moats and for ramparts for defence. It is virtually impossible for us, in today's highly technological society, to have any realistic concept of the scale of manual labour and the unremitting endeavour required for such works. Until the steam engine was developed during the nineteenth century, all excavation was not only done by hand but with the most primitive of tools: antler picks, stone axes, followed by the improved digging tools of bronze and iron. But human muscle powered them all. That was a feasible proposition when slave labour, or almost the equivalent, in unlimited numbers was available and time was of little consequence. The Chinese military engineers who built the 2,000-mile (3,200-km) Great Wall of China – which remains the only single man-made structure visible from space – probably had an open-ended completion date with the leeway of a century or so either way.

Before the Industrial Revolution, a man could begin work as a youth on a major project which, even in old age, he would not live to see completed. The great Gothic cathedrals of the eleventh century, for example, took an average of a hundred years to complete with generations of masons, carpenters and labourers spending their entire working lives on the construction of each one. They were, however, altruistically built 'to the glory of God'. Time was of no consequence; no additional indulgence would be gained if the building were to be finished twenty years or so sooner than expected. There was piety enough in the very act of construction. Closer to our own time, the patronage of a noble family in engaging a landscape gardener such as Capability Brown to create a vista of a wooded hillside or to excavate 20 acres (8ha) for a lake in the parkland of the estate would also be a lengthy undertaking, requiring at least two years of continuous manual labour from more than 500 men. Since the completed landscape

Detail of the 13,000-ton walking dragline excavator, the Bucyrus-Erie 4250W 'Big Muskie'. This massive machine can scoop up 220cu yd (300 tons) of spoil with each pass of the bucket.

would not be at its best for another fifty years or so, the time taken in construction using picks, spades and horse-drawn wagons was neither considered nor questioned.

With the coming of the Industrial Revolution, that most agreeable *laissez-faire* attitude to time ended. The end came with the newly created railways. They had to have an agreed universal time, valid throughout the country, for their timetables. Before the railways, bells had summoned children to school and the pious to church. The railways demanded clocks, and accurate ones. As industry grew, time became money. A ship lingering on the stocks, for example, was not earning revenue. Steam-driven earth-moving machines, as they increasingly became available, were used. This was not because they were cheaper than manual labour (which, in nineteenth-century Britain was both cheap and plentiful) but because machines had one overwhelming advantage: they could perform a given task far more quickly than gangs of low-paid labourers, however large. But it was to be some time before this fact was recognized in Britain.

As Britain is an island perhaps it is not surprising that the first known machine used for digging was a steam-powered excavator, mounted on a barge and put to work as a dredger maintaining shipping channels in the shallow estuaries of rivers. The excavator was built on Tyneside by the Grimshaw Company in 1796. Apart from the fact that it was known to be steam powered, very little else is certain. No drawings exist and there is no record as to how well, or otherwise, the Grimshaw machine worked. It is interesting to note that this pioneering, though ephemeral, essay into powered excavation should have been the one aspect of digging that men, however plentiful and inexpensive, could not undertake: underwater dredging.

The early railways (Stevenson and his contemporaries would have used the term 'railroads', still current in the USA; the word 'railway' was coined to denote a legal right of way as sanctioned by a Parliamentary Act) were, like the British canal system of the eighteenth century, largely built by manual labour, often Irish. This might seem strange, because the railways began as steam powered and remained so for the next 120 years. The answer is simply that from the 1840s when 'railway mania' caused the rapid countrywide spread of the railways in Britain, there were insufficient steam-powered excavators in existence to make any significant impact on the construction. Even if there had been, the problems of getting them to the sites would have been insuperable on the roads of the time, the lack of roads suitable for heavy freight being a major justification for the building of the railways in the first place. Finally, there was the question of cost. As has been said, labour was both plentiful and cheap, but earth-moving machinery was neither. The time factor was not as pressing as might be supposed: earth moving by the tens of thousands of 'navigators' or 'navvies' was only part of the construction of the railways. There was also the extensive building of bridges, retaining walls, tunnels and track laying as well as the stations and other buildings that were inseparable from a railway system and for which the steam-powered machinery then developed would have been useless.

In the United States, the situation was different. There, manual labour, despite the waves of immigration, was the reverse of the situation in Britain, being neither cheap nor plentiful. Indeed the US railroads were built with the help of large numbers of Chinese 'coolies', as they were known, imported for the task. The chronic lack of available labourers for manual earth shifting led to the United States becoming the main source of steam excavators, a lead they have never lost.

Credit for the first practical land-based steam excavator belongs to a young American, William Smith Otis, whose father was a partner in a firm of Philadelphia contractors, Carmichael, Fairbanks & Otis. William Otis, in 1834, commissioned the Philadelphia engineering company of Eastwick & Harrison to build to his designs an 'Otis Steam Shovel'. This, the first steam-powered shovel, was mounted on a rolling chassis which had four standard-gauge railway wheels, enabling it to operate on railway tracks. From a line drawing, which later appeared in the prestigious *Proceedings of the Institution of Civil Engineers*, published in London in 1845, one can see that the Otis steam shovel looked remarkably similar to later steam excavators. It had salient features which were to become standard. A vertical steam boiler and a single reversible steam engine were placed to the rear of the chassis in the charge of a fireman. To the front, a cast-iron pillar about 10ft (3m) high supported a triangular wooden jib which could be slewed by the engine for 90 degrees from the centre line to left or right. The large bucket or shovel was pivoted from the jib with a racking gear to set the depth of the cut and was able to be moved, by a chain under power, forward and up, thus mimicking the action of a navvy's shovel, only with ten times the capacity. Having 'dug' the load, a cranesman standing by the jib controlled the machine and could raise the shovel to retain the spoil it contained, slew the jib until it was over a wagon or tip, and then release the rear door of the shovel to drop the load by gravity. The Otis appears to have had a self-propelling drive available but one supposes that this was for moving while operating rather than an '*en-route*' capability.

The prototype Otis steam shovel was used on the construction of the trackbed of the Baltimore & Ohio Railroad in 1834. It must have met with approval because William Otis, in 1836, applied for and was granted US patents for an improved design of an 'American Steam Excavator'. This is possibly the first time the word 'excavator' was applied in that context. Eastwick & Harrison built at least four machines to the developed Otis design. One was used in 1838 in the construction of the Boston & Albany Railroad; two, rather surprisingly, went to Russia and were never heard of again, and the fourth came to England for use in the building of the Eastern Counties Railway. This machine was illustrated in the *Proceedings of the Institution of Civil Engineers* of 1845. William S. Otis sadly did not live to enjoy the fruits of his invention; he died tragically at the early age of 26 in 1842, killed, so it is said, in an accident involving one of his machines.

The single Otis excavator imported into England was to make a profound impression on British engineers, who lost no time in designing remarkably similar machines. It was to be the start of the excavator industry in Britain, aided, it must be said, by the convenient expiry of the Otis US patents in 1860. James Dunbar patented his version of the Otis design in 1874, calling it the 'Dunbar Steam Crane Navvy'. The most original aspect of the patent would appear to be the use of the word 'navvy'; steam navvies became, in Britain, the generic term for all excavators, remaining so until the 1950s.

Dunbar contracted the well-known Lincoln engineering company, Ruston, Proctor & Co., to construct his steam shovel. The company had the acumen to see the promising sales potential of the design and forthwith bought up all of the Dunbar patents. The prototype 'Ruston-Dunbar Steam Shovel' was completed in 1875 and immediately sold to a firm of public works contractors, Lucas & Aired. Other orders followed, and hundreds were to be sold in the next few years. They had appeared at precisely the right time. The Albert Docks in London were being excavated between 1875 and 1880 and, perhaps the most important British civil engineering project to date, the Manchester Ship Canal was being constructed. The excavation of this canal, starting in 1887, employed no fewer than seventy Ruston-Dunbar steam shovels, by far the greatest number to be used on a single project up to that time. The Ruston machines had cornered the market, to the disadvantage of all competitors, though one, the Whitaker Steam Shovel Company of Leeds, had perfected a 360-degree slewing system. This amounted to placing the entire excavator, boiler, engine, jib and bucket, on a base that was in effect a turntable mounted on a railway-gauge flat-bed truck – a practical layout that was to be much copied. Ruston's strength lay, first, in the holding of the Dunbar patents and, second, in the fact that the company was in a strong position as manufacturers because they had 'off-the-shelf' Ruston steam engines ready to power their steam excavators.

The Victorian age in Britain was one of relentless, confident progress. The building of the Albert Docks in London and the Manchester Ship Canal were only the beginning of this. The railways, after the first flush of hurried construction, were consolidating into the definitive national network that was to survive until the swingeing cut-backs instigated by Dr Beeching in the 1960s. There were new docks to be built in most ports, improved metalled roads, thousands of miles of municipal sewers and reservoirs, gas mains and countless civil engineering schemes, all of which required millions of cubic yards (cubic metres) of earth to be moved. The steam shovel was the only practical mechanical aid for a still remaining vast army of manual labourers, the original navvies, most of whom resented and feared those 'new-fangled American devils' – the tireless steam excavators. Since the Ruston machines were a direct development of the original Otis, it may be of interest to know that the prototype Otis survived for the next seventy years – doubtless much rebuilt – in full working order. Sadly, it was broken up in 1905, a remarkable record for a machine made very largely from wood.

As the Ruston excavator of 1874 was the first commercially successful British design, it is worth looking at the construction in some detail. In its general arrangement the Ruston clearly is a development of the Otis design. It could hardly be otherwise, as the Otis excavator, like Bell's threshing-machine, defined the essential basic functions so elegantly that, the ability to slew through 360 degrees apart, few, if any, alternative general arrangements existed. Certainly none have as yet appeared. Both designs were extensively to be developed and powered by steam, electricity and diesel, but were never superseded. The fundamental difference between the Ruston and the Otis lay in the fact that the Ruston was mainly of steel construction. It was mounted, as was the Otis, on a four-wheel standard-railway-gauge flat-bed base with a drive to the front axle for limited movement along the rails while working. The steam vertical boiler stood, as in the Otis machine, to the rear of the base. The two-cylinder steam engine of conventional design, curiously non-reversing, was mounted in front of the boiler. The engine powered all the functions of the excavator through gearing engaged either by dog or friction clutches. The most striking feature of the Ruston was the very substantial triangular structure that supported the steel lattice jib. The bucket arm, which was a wooden beam cladded in steel, was a form of structure empirically found to be best suited to the shock loadings imposed by the action of the digging, and a method of excavator bucket-arm construction that would remain standard for many years to come.

One of the earliest names in excavators is Bucyrus. The American firm supplied seventy-seven of their steam-driven 70- and 90-ton machines to excavate the Panama Canal, perhaps the greatest civil-engineering feat of the twentieth century. Here, two of the Bucyrus excavators make the final, historic cut to complete the Canal, c.1909.

The bucket arm could be extended or retracted to adjust the depth of cut of the bucket manually by the cranesman turning a large hand-wheel, which operated racking gears via a long chain-driven shaft. The hand operation of the racking was thought to give the cranesman finer control of the digging bucket than the Otis engine-powered racking. The bucket arm could be slewed left or right, powered from the engine by chains on slewing drums engaged, as required, by the cranesman through friction clutches. A friction clutch also engaged a chain drive to the railway wheel axle to move the excavator along a railway track when working jacks on outriggers were fitted to stabilize the machine, vital when the 2-cu yd (1.5-cu m) capacity bucket was fully loaded and the arm was at maximum extension and fully slewed.

The steam engine powered the bucket arm via a hoisting drum through a dog clutch and forward and reverse gearing with a pull of up to 10 tons. Chains were used on the early machines as chains were found better to be able to withstand the strains imposed by the slewing action of the bucket arm than wire cable (the drawback in the use of chains is that, unlike wire cable, chains give no warning of impending failure which, with men working around and, at times under, the excavator bucket was clearly a hazard). The all-up weight of a Ruston excavation of 1874 in working order was 32 tons, which aided stability on the narrow base imposed by the railway's restrictive loading gauge.

A substantial number of the Dunbar & Ruston excavators were made. They became a familiar sight in the London of 1894 to 1899 as over forty of them were in use when the Great Central Line was extended into the capital to the last-built main-line London terminus, Marylebone. Because the Otis/Ruston design of excavator being used was exclusively mounted on a railway-truck base, moved on railway lines and was working extensively on railway projects, the type came to be known, both in Britain and the United States as 'The Railroad Type Shovel'. In the United States size is important. The 'Railroad Type' was to outgrow the simple restricting four-wheel railroad truck and very much enlarged versions, up to 90 tons, appeared on multiple bogies, the vertical boiler being superseded by a larger, more efficient horizontal locomotive type. As the work loads increased, separate engines were to be provided for the disparate functions of hoisting, slewing and racking.

New names appeared. The Otis family, wishing to continue in the excavator business after William Otis's untimely death, began to look around for suitable partners. William's widow remarried, her new husband being Oliver Chapman, and the steam shovels were marketed under the name Otis-Chapman. This was to change again when John Souther joined the firm. His contribution must have been considered significant as the machines were now called Chapman-Souther; they enjoyed moderate success up to 1880, when they were to be overshadowed by upstart newcomers.

It is a curious fact that the makers of steam shovels in the United States seemed, after the demise of the Otis, to call their products after the name of the town in which they were produced. This might be because in the United States, at the turn of the century, excavators were built by companies that were run by partnerships, rather than a single autocratic boss like Henry Ford. Perhaps unable or unwilling to upset the collective management by singling out one name, they opted for place names instead. Whatever the reason, the two US makers of the excavators that were to dominate the international market for a century adopted place names for their products: Marion and Bucyrus.

The Marion Steam Shovel Company came into existence in 1884. It had been formed in the town of Marion, Ohio, by a steam excavator operator named Henry Barnhart who decided that he could, with expert engineering assistance, produce a better machine than those he was using (possibly Chapman-Souther). He joined George King and Edward Huber, the latter owning an engineering plant which made agricultural machinery in the town of Marion. By 1900 the company was producing a range of excellent steam shovels. They had left the constraints of the railroad mounted excavator behind. The 'Marion Portable Steam Shovel' was mounted on a sturdy steel underframe with wide-tracked and treaded-steel wheels supporting a 360-degree slew turntable on which the 'works' were mounted. The Portable Marion was claimed to be the first 360-degree slewing excavator offered in the United States. Twenty-four Marion steam shovels were put to use in the digging of the Panama Canal, an enormous undertaking of civil engineering, which made the Manchester Ship Canal seem a mere culvert by comparison.

The Panama Canal aside, from 1911 the Marion Steam Shovel Company also sold a 'stripping' version of their excavators, the Marion Model 250, which was widely used in the growing number of open-cast mineral mines, a form of mining that was pioneered in the United States. Substantial orders for strippers came to Marion from Canada, and two were exported to England to work in the limestone quarries of the Midlands.

Steam stripping shovels were essentially a taller, longer, jib-arm version of the standard excavator. Their function was not so much to dig deeply but to 'strip' the overburden covering the ore and coal deposits from the steep sides of a quarry. The Marion 250 stripper was rail-mounted but not on standard-gauge track; short lengths of rail with a gauge of around 12ft (3.6m) were laid as required, allowing the stripper to move as it worked the adjacent quarry faces. Since they had farther to reach, the jibs of strippers were longer and lighter built; at least twice the length of the equivalently sized steam shovel and much more like the jib of a conventional lifting crane. The stripper bucket arm was also longer and lighter and had what was called a 'stiff leg' or 'knee' action. The bucket, called a 'dipper' on strippers, differed too, designed to penetrate the overburden deeply while not overloading the lifting capacity of the long bucket arm. The rail track soon disappeared, being replaced by a development known as 'four-corner self-levelling gear'. The first was the Marion Model 36-E excavator with the provision of four separate crawler tracks to support and move the machine.

During fifty years of production of stripping shovels, the Marion company offered larger and larger machines, from the 18-cu yd (13.7-cu m) 5560 of 1932 to the 60-cu yd (45.9-cu m) 5760 of 1956. In 1965 Marion had the distinction of building a truly awesome machine: the Marion 6360 stripping shovel which weighed in at an unbelievable 12,600 tons. By some margin it was the largest ever built or ever likely to be built. The 180-cu yd (137.7-cu m) capacity dipper bucket of the 6360 would have accommodated a London bus. Other statistics of the 6360, nicknamed 'The Captain' by the men who worked it in the Captain mine at Percy, Illinois, were also impressive. It was supported on no fewer than 8 crawler tracks, each of which stood 16ft (4.8m) high and 44ft (13.2m) in length. The total horsepower of the machine was 33,000, the jib or boom was 215ft (64.5m) long and the hoist cables, 4 in double sheaves, each had a diameter of 3.6in (8.75cm). The past tense is used because the massive record-breaking Marion 6360 was destroyed by fire in 1991. Although the 6360 remains the largest of the classic excavators and strippers built, there were serious competitors along the way, principally the second American manufacturer, Bucyrus.

Bucyrus is the name of a small town in Northern Ohio, on the Sandusky river and 45 miles (72km) south of Lake Erie. The company, always known as Bucyrus, had been a producer of railroad equipment. It manufactured its first steam shovel in 1882 for construction work for a railroad company, probably the famous Baltimore & Ohio. The excavator must have been found to be satisfactory because, from 1893, the company, still named Bucyrus though it was now situated in South Milwaukee, Wisconsin, had within twelve years sold 170 Bucyrus steam shovels. These were working on civil engineering contracts. The Chicago Drainage Channel employed twenty-four Bucyrus shovels and the company's share of the Panama Canal project was major: from 1908 no fewer than seventy-seven Bucyrus 70- and 90-ton excavators worked alongside twenty-four Marions. Bucyrus excavators were also extensively used in open-cast mines at home and abroad. Several were exported to Spain to work in copper mines and others to operate in the limestone pits of the English Midlands. Bucyrus machines were built under licence in Russia from 1900 and in Canada from 1903.

One of the few surviving competitors to the Marion-Bucyrus near-monopoly was the Erie Steam Shovel Company of Pennsylvania. This was acquired in 1927 by Bucyrus, the company becoming Bucyrus-Erie, with a wide range of excavators and strippers being produced. Although the Marion 6360 stripper can claim to be the biggest ever, Bucyrus-Erie could also produce very large machines; the Bucyrus-Erie 3850-B had a 115-cu yd (87.9-cu m) dipper bucket. Another, 'The River King', which worked the River King Mine at Marissa, Illinois, weighed 8,800 tons and had a massive 140-cu yd (107-cu m) bucket. This machine worked continuously in the same mine for twenty-eight years, from 1964 to 1992, during which time it had travelled 20 miles (32km) digging all the way. At the end of the rift this splendid machine was sadly cut up where it stood for scrap, all too often the fate of old excavators which, having been erected on site, were too big and too expensive to remove, overhaul and be erected elsewhere.

The British connection with excavators began, as we have seen, with what were frankly copies, official and otherwise, of the early American designs. It would become tedious to define the many ephemeral attempts to compete with the American imports. Whitaker of Leeds, as noted, designed the first fully slewing steam shovel. Priestman Brothers of Hull was formed in 1874 by Samuel and William Priestman to build steam-operated 'grabs'. Later, the company offered 'grabbing' cranes which were successful, including one design of the 1920s mounted on a tracked ex-WD gun carriage and another which, on the evidence of a surviving photograph, seems to be fitted to a tank's caterpillar tracks; this may be the machine known to have been built for assessment by the army. If so, it was to be cut up very shortly after the army trials. Although Priestman offered the complete cranes, they were essentially the makers of grabbing buckets which could, with little modification, be fitted to any existing crane. Grabbing and dragline Priestman excavators were used very successfully dredging rivers and other inland waterways. The Priestman No. 5 excavator was to remain in production for ten years from 1924; it was then redesigned to become the Priestman Cub and was to be proved to be a most versatile machine, able to be fitted to act as skimmer, crane, dragline or shovel, and to remain in production until the late 1960s.

The Priestman company had a formidable competitor in another English firm, Rustons of Lincoln, which had successfully developed their machines from 1875 and by 1918 had formed a partnership with the Hornsby Company, becoming Ruston-Hornsby. By 1920 the firm had advanced to the Ruston-Hornsby No. 4 excavator, which was a commercial success. However, with the stock-market crash of 1929 and the depression which followed, the prospects were looking bleak with, at best, a very restricted market for excavators in view. On top of that, the company had faced up to the inevitable conclusion that the American excavators being imported into Europe were in many ways superior to home products, including their own. Knowing that Bucyrus were keen to secure a foothold in Britain with its worldwide preferential Empire markets, Rustons approached the Milwaukee company and found that they were pushing against an open door. In January 1930 Ruston-Hornsby amalgamated with the American Bucyrus Company to become Ruston-Bucyrus, a name which, when first heard in England, suggested some rare, horned animal discovered by a zoologist called Ruston in some distant, obscure Ural mountain range, to be equated with Przewalski's Horse or Thompson's Gazelle. The excavators had a certain animal-like movement, and became a common sight throughout British construction between the wars, as they bobbed and turned, digging for the new arterial main roads and extensive new suburban housing estates.

Most of the Ruston-Bucyrus excavators built were the classic bucket type but grabbing excavators were also built by the Lincoln factory. The Ruston-Bucyrus No. 4 was a grabbing excavator; 934 were produced from 1926 to 1933. Only three are known to have survived; there is one in New Zealand, a second in the Museum of Lincolnshire Life, and the third is privately owned by Mike Bent-Marshall, the son of the original owner. This Ruston-Bucyrus No. 4 has now been restored. It is of interest, therefore, to look at this machine as it is representative of a large generation of medium-sized excavators, built and used widely between the wars. The Ruston-Bucyrus was delivered new as one of five ordered by the Essex earth-plant contractors, Bent-Marshall, from the Lincoln works in 1931 at a cost of £1,400 which was a very considerable sum in those days. It worked from 1931 until 1959, after which it was relegated to function as a yard crane. The Ruston-Bucyrus No. 4 was offered powered by a number of alternative engines including a twin-cylinder steam engine, a four-cylinder petrol/paraffin engine and a twin-cylinder diesel engine. The Bent-Marshall machine had the original engine replaced in 1956 by a 32hp Dorman two-cylinder diesel. The excavator has a substantial crawler tracked drive and a 360-degree revolving upper frame which contains the controls and all the lifting and slewing gears. Photographs taken during the painstaking restoration give a very clear idea of the sturdy machinery and good engineering built into the excavator. Since the restoration, this machine is regularly shown at the annual Essex vintage machinery exhibition.

The Ruston-Bucyrus No. 4 was, as already mentioned, a specialized form of excavator known as the dragline type. The virtue of the dragline machine is that it can excavate at a far greater range than the jib and boom machine. In the simplest terms it is a crane with a bucket, the open end of which faces the excavator and is attached to the jib by a normal hawser. Another hawser, the dragline, is attached to the front of the bucket by which it is dragged towards the crane with a pull of several tons, filling with spoil as it does so. At the end of travel the driver engages the hoist control, the full bucket is lifted clear and the machine is slewed to drop the spoil on to a waiting truck or wagon. The dragline excavator, therefore, works in the opposite sense to the conventional bucket-arm type in that the bucket is dragged towards the excavator, whereas the jib and boom shovel machine pushes the bucket or dipper away from itself. Dragline excavation is extensively employed by drainage and harbour boards as it is particularly suited to working for dredging docks, ditches, dykes, culverts, rivers and lakes since, unlike a conventional

A Ruston-Bucyrus RB10 excavator restored by Ian Hartland (*pictured*). The RB10 was a very popular machine during the 1930s. Though small by later standards, they fulfilled a need, and many were working in the early British open-cast coal mines until bigger machines became available from the United States after the Second World War.

excavator, the dragline can excavate at a level well below itself. When dredging, for example, the lifting hawser can be released at the maximum distance allowed by the length of the crane jib to drop the bucket, which sinks under its own weight to the bottom of the river or waterway being dredged. Dragline excavation was pioneered in Europe by the Ruston firm using a crane jib and bucket. Before that an earlier form of dragline excavation was the first to be performed by mechanical means. A pair of steam ploughing engines worked on opposite sides of a lake or river, dragging a reversible bucket or scraper backwards and forwards across the lake bed to dredge it of weeds and silt. Even to this day, preserved ploughing engines are occasionally still used to perform this service but the dedicated dragline excavator is the preferred method.

Ruston-Hornsby, before they amalgamated with Bucyrus, were very much involved in some major dragline excavations. In 1923, a number of Ruston-Hornsby excavators, at the time among the largest in existence, were being used to dig irrigation canals for the Indian government. There were several schemes, one being in the Sind province of Sukkur where an extensive system of canals some 200ft (60m) wide and 12ft (3.6m) in depth were to be dug to irrigate over 7 million acres (2,833,000ha). The Ruston machines shipped out from Britain had to be erected by the company's engineers on site. Each excavator consisted of over 1,000 parts, the heaviest weighing 19 tons, all to be assembled in conditions of extreme heat, bad water and fever. One Ruston engineer was bitten by a snake while he was oiling the mechanism. He survived – just. The massive 350-ton dragline excavators were steam-powered on wide-track rail mountings and were big by the standards of the day. The jibs were 120ft (36m) long and the dipper buckets had a cutting power of 30 tons which could excavate 10 tons of earth at a cut, a work rate that enabled each single Ruston-Hornsby dragline to fill a sixty-wagon train every hour. The cycle of operation of running out the bucket, dropping it at the start of the cut, dragging, lifting and unloading took, in the light earth of the Indian sites, under a minute. On average, at one site in the Sutlej valley, each Ruston-Hornsby dragline could excavate nearly 250,000cu yd (191,250cu m) every working day, equivalent to the work of 8,000 local labourers. The main dragline engine of the Ruston-Hornsby machines was of 400hp and the slewing engines were rated at 200hp. All the engines were steam powered. The coal bunkers on each machine held 4 tons of coal and each had a steam-powered hoist to keep them trimmed. As a crane the machines could lift 22 tons at 125-ft (37.5-m) radius. The Ruston-Hornsby excavators were so successful that at least a dozen were ordered. It was possibly this market in the British Empire that had attracted the attention of Bucyrus. The Indian projects were still continuing after the merger in 1930.

Dragline excavators move using either rails or crawler tracks. These methods are satisfactory when the ground of the quarry or mine is reasonably firm; if, on the other hand, the going is soft then the need arises for an alternative system – the 'walking dragline'. Walking dragline machines originated in the United States, invented in 1904 by John Page, a partner in a company contracted to excavate a canal in Illinois. Page probably used a modified crane but soon dedicated machines appeared. One of the first was produced by the Monaghan Company in 1913. This machine dispensed with rails and caterpillar tracks and used specially designed 'shoes' to become a true walking dragline. In simplistic terms, the excavator sits on a solid base or 'tub', as it is known. This tub is in contact with the ground and provides a secure foundation for the roller path, which allows the entire structure to slew through 360 degrees when dragging and dumping. When it is required to advance to a fresh cut the 'shoes', powered almost universally by electrical power, are moved by cams in an eccentric, circular motion, lifting the machine clear of the ground and lowering it a given distance forward. Because the shoes are very wide and long, the footprint weight they exert on even very soft ground is less than that of crawler tracks. Although the actual distance covered by each 'step' is not large, the total distance covered by some walking draglines is surprising: one British Ransome-Rapier excavator walked, in 1940, a total distance of 13 miles (20.8km).

Since the end of the war in 1945 there have been major advances in the design and construction of excavators. Steam power has all but disappeared; a few steam-driven machines linger on in the Third World and others are preserved in working order by steam enthusiasts. Today, diesel or diesel-electric power is favoured. The size of the machines has grown as they tend to be erected in the quarry or open-cast coal or ore mines in which they will spend their working lives; there are no restrictions imposed by the need to travel on public highways or railways.

The main contemporary makers of walking draglines are American, Russian and British. The Marion Company, which had constructed the world's largest stripper shovel, has also made some of the largest dragline excavators: one, a Marion 7400-M, worked in an iron-ore mine near Scunthorpe. The British company of Ransome & Rapier has produced some impressive walking dragline excavators; their W-1800s have buckets with a capacity of 33cu yd (25.2cu m). One W-1800, working in an anthracite open-cast mine in South Wales, was claimed to be the largest in Europe; the W-1800 had a 40-cu yd (30.6-cu m) bucket fitted and operated from a 247-ft (74-m) boom. When ordered new in 1961 it cost nearly £1 million but worked twenty-four hours a day non-stop. Eventually, at the end of the contract, it was dismantled, overhauled and sold to an American mining company at a considerable profit.

The 'world's largest dragline' title went firstly to 'Big Georgie', a Bucyrus-Erie 1550-W, which could accommodate two large saloon cars side by side in its bucket. However, this machine had a challenger for the world title from the 'Ace of Spades' made by P&H, a German/American firm. British Coal bought the machine in 1989; it took 18 months to construct from June 1990 to December 1991. (The name, incidentally, was open to local competition and a small girl came up with 'Ace of Spades'.) The dragline remains the largest in Europe, working twenty-four a day apart from half an hour off each Tuesday for routine maintenance. It cost £12 million to buy but it has a thirty-year life expectancy and should still be working well into the 2020s – which is just as well, because there are an estimated 12 millions tons of coal yet to be excavated in the open-cast mine at Stobswood where the Ace of Spades works.

There are some very big walking draglines in Russia, some weighing in excess of 10,000 tons. But, as is usually the case when size is the subject, the United States is the leader. The undisputed heavyweight of the walking dragline world is now the Bucyrus-Erie 4250-W. This machine has a bucket that can take a bite of nearly 300 tons (220cu yd/168cu m) suspended from a 265-ft (79.5-m) boom. The 4250-W has an all-up weight of 12,244 tons, which requires some 45,000hp of electrical energy to operate. It is no greyhound; it 'walks' at only 0.16mph (0.25km/h) but with over 12,000 tons on the move it must be an awesome sight.

Although the mammoths seen at work in open-cast mines around the world are without doubt the largest ever likely to be built, by far the largest number of excavators at work are

The title of the largest excavator in Europe is now bestowed on the 'Ace of Spades', a P&H 757 dragline working at Stobswood, Northumberland. At £12 million, it is also probably the most expensive excavator working in the UK.

small enough to fit four or more into the buckets of machines like the Bucyrus-Erie. These are the yellow-painted JCBs which dominate building and construction sites. These versatile machines are offered in a bewildering range; from the mini-excavators slim enough at 3ft 3in (1m) to be able to squeeze through a garden gateway to the powerful and compact backhoe loaders and tracked excavators. The JCB range relies on hydraulic power to operate.

Hydraulically operated excavators are as old as any of the others. First experiments using fluids under pressure to transmit power through pipes to rams were instigated by Sir W.G. Armstrong as long ago as 1880; other work followed and several excavators appeared tentatively using hydraulic transmission to replace certain functions hitherto carried out by ropes, cables, chains or gears. It was after the end of the war in 1945 that engineers, drawing on wartime experience gained with hydraulically operated military aircraft controls (bomb doors, engine throttles, flaps, wheel brakes, etc.) began the application of hydraulics to excavators which were to play a significant role. It was the development of compact though powerful hydraulic motors, pumps and reinforced flexible piping able to withstand very high pressures in a harsh environment which was the important advance.

From 1960 a range of long-reach excavators with the old winding drums and cables replaced by hydraulic pumps and motors appeared on the market. The old-established British firm of Priestman of Hull began the trend in Europe while, in the United States in 1970 a new name, Koehring, offered extra-long booms on a range of hydraulic excavators and demonstrated that, in the small to medium sizes (the range used by building and urban civil engineering) the hydraulic operation of an excavator with a bucket capacity of up to 3cu yd (2.3cu m) was superior to conventional cable operation because the boom length was not restricted by the stretch and sag and inevitable loss of precision a long cable would incur. The frictional losses in a hydraulic system were also low. In short, the hydraulic excavator could perform better than any equivalently sized cable-operated machine. In the main, hydraulic operation offered a longer reach, greater efficiency, was much faster and far more precise in operation. In the smaller range the early machines were tractor mounted and the Whitlock Dinkum-Digger series of digger-loaders were popular on the farm. The Poclain Company built small excavator/shovels mounted on conventional rubber-tyred wheels. However, the market was to be dominated by the Uttoxeter-based JCB company.

The letters JCB stand for Joseph Cyril Bamford, one of a vanishing breed of men who, in 1945, made his first product with his own hands. This was a farm trailer built from metal salvaged from old air-raid shelters. The trailer was sold for £45, and from this was created a major international company. Other trailers followed, significantly with hydraulic tipping. Next, a hydraulically operated forward loader fitted to Fordson tractors was built and sold and, in 1954, the first of what was to prove a very long line indeed of hydraulic excavators appeared. Joe Bamford had commissioned his chief designer, Alec Kelly, to design a small hydraulic excavator built around a Fordson Major tractor with the needs of farmers very much to the fore. The sales leaflet of that first JCB excavator lists the specifications: it had a 180-degree slew arc, because farmers had asked for a machine that could run parallel to a ditch so that it could be dredged clean. In fact, the prototype excavator had already been built with only the usual 90-degree slew; it was only after talking to farmers at a local agricultural show that the ditch-dredging capability was made known to Joe Bamford. He rang the works from the show and instructed Kelly to discard the prototype and start again with the 180-degree slew. Listening to the needs of ordinary customers is obvious but is seldom actually put into practice by British firms. JCB was, from the start, an exception. Development went on apace. There were technical problems with the hydraulic transmissions, cavitation being the main difficulty; these were overcome and, by 1957, over 200 Mark 1 JCB excavators were sold. One, working in a Derby coalyard, was not to be replaced for thirteen years. The next model was the Hydra-Digger, and over 2,000 of these were sold. The definitive JCB, the bright yellow-painted, now-familiar, 'backhoe-bucket' excavator appeared in 1960 as the JCB 4. This machine had, for the first time, a simple two-

This dragline excavator, a Bucyrus-Erie 1550-W can claim two records when it worked in an open-cast coal mine in Northumberland in 1970. It walked $2\frac{1}{2}$ miles from one site to another, crossing three roads, a railway line and a river. At 3,000 tons, this must have been a sight to remember. At the time, the 1550-W was the largest excavator in Europe.

lever control, which meant that the operator had the machine's entire function under his two hands without having to change hands as on the earlier, five-lever Hydra-Digger.

The JCB range, both today and forward into the twenty-first century, is comprehensive. There are six mini-excavators from 1.4 to 3.6 tons, two wheeled excavators in the 13- to 17-ton class and no fewer than twelve tracked excavators with capacities ranging from 7 to 45 tons. There are eight versions of the most visible of the JCB range, the popular backhoe loaders seen in practically every works yard and building site. Indeed, as all vacuum cleaners tend to be called Hoovers, so all small excavators and loaders are thought to be JCBs. Most of them are.

There is a final form of excavator that does not fit in with the many that have gone before. From the Otis of 1831 to the colossal dragline machines of today, all have a common feature: a single bucket which is either dragged or pushed to gather material either to dig a hole or to collect minerals. Over the years the difference is basically that of scale. The 'continuous excavator' is different. In place of the single bucket, the continuous excavator works like the shipping dredger, with an endless loop of buckets, like the steps of an escalator, continuously on the move.

Steam-powered dredgers were put to work in the construction of harbours and the creation and maintenance of deep-water channels in rivers and estuaries from the middle of the nineteenth century. They still perform the same duties today. Towards the end of the nineteenth century the endless bucket principle was applied to dry land for open-cast mining of iron ore, coal, and clays for brick-making. Like the conventional single-bucket excavator, the endless-bucket type was mounted on rails and was powered by a steam engine, though electrical operation was universal by the 1920s. The market leader was, from the start, a German firm: Backau-Wolf. The first machines simply emptied their buckets by gravity as they ran over the top of the loop, usually into a chute placed above railway wagons or on to a conveyor belt for disposal. In 1927 the German company developed their machines to have a slewing boom supporting the bucket chain, the revised excavator having a much-increased capacity. By 1939 Backau-Wolf, the market leader, had constructed the largest bucket-chain excavator in the world. It weighed 2,500 tons and each of the thirty-eight buckets in the chain had a capacity of 333gal (1,500l).

There are variations on the endless bucket theme; the ladder type is possibly the most used and it has certainly produced some of the largest. Again developed and constructed principally by German companies, many of these machines weigh well over 12,000 tons. In a sense, the ladder type is a very high-capacity dragline excavator. In place of the single bucket being drawn to the machine by a cable, the ladder machines suspends, from a lattice boom, a second and longer boom. This is the 'ladder', which has the bucket loop continuously travelling around it. The buckets on the lower side of the ladder are in contact with the mine surface, are drawn towards the excavator and filled up with the ore or minerals being mined – hence the dragline analogy. When a bucket reaches the near end of the ladder boom it is up-ended as the direction of travel changes and the contents fall on to conveyor belts which, in a big mine, might travel for miles to the collection point. Even as late as the 1960s, these huge machines were rail mounted, often on twin standard-gauge railway tracks. The suspension boom can be raised or lowered by cables and sheaves so that the machine can excavate above the track on which the machine moves, below it, or level with it. One of the latest bucket

Left Not all excavators are massive. This very early JCB Mk I excavator, built around a Fordson Major tractor, was all hydraulic and has been continuously at work for forty-one years on the Harper Bower Farm. Howard Wood is at the controls of what JCB think is the earliest surviving Mk I excavator.

excavators, the Krupps RS 1000, has dispensed with railway lines; instead it has a complex system of very large crawler tracks, allowing for greater freedom of movement.

There is a third variant of the continuous bucket principle: the wheel excavator. In this system the buckets, typically twelve, are mounted as a wheel on the end of a long boom which is lowered to contact the mine surface and scoop up the minerals in the same way as the ladder type. The size range of bucket-wheel excavators is very wide, from small one-man operated machines up to 12,000-tonners. The German company of Orenstein & Kopple (O.K.), Krupp, Siemens, and others, offer wheel excavators. In Britain both the old-established firms of Ruston-Hornsby and Ruston-Bucyrus have made them, as have American and Russian concerns. They are at work all over the world, mining brown coal, lignite and bauxite. On a smaller scale, one was used to enlarge a Buckinghamshire reservoir while another is working for a cement manufacturer in Kent.

Finally, a most unusual and early Ruston-Bucyrus wheel excavator was involved in a highly classified operation during the Second World War. It bore the code-name 'Nellie'. In late 1939, when the 'phoney war' was at its most serene, Winston Churchill, not yet Prime Minister but First Lord of the Admiralty, came to the conclusion that the German Siegfried line, far from being a convenient place on which to hang washing, as the popular song of the moment suggested, did in fact present a most formidable obstacle to the invasion of Germany. This flawed conclusion was drawn before it had been shown, by the German army in May 1940, that any defensive system of forts, however extensive and apparently formidable, can be readily by-passed, as was the supposedly impregnable French Maginot line. Churchill, drawing on his trench warfare experience of the First World War (as did most British and French generals of the time), decided that a frontal attack on the Siegfried line would be suicidal unless some form of cover for the attacking troops and artillery could be provided. Churchill thought he had the answer and, being the First Lord of the Admiralty, it was appropriate that he sent for the Director of Naval Construction, Stanley Goodall. He asked Goodall to create a design forthwith for a machine to be employed in a frontal assault on the Siegfried line. Just what Stanley Goodall thought of this proposition

Overleaf When it comes to size, the contemporary 'bucket-wheel continuous' (BWE) excavators are the largest man-made earth-moving machines in existence. This German monster, the Orenstein & Koppel (O&K) 289, weighs in at 14,028 tons and has a capacity of 314,000cu yd (240,000cu m). Even larger BWEs are planned.

KRUPP
SIEMENS

'Nellie' performs for Winston Churchill. in 1940. The Prime Minister ordered several of these unconventional bucket-wheel excavators for a projected attack on the German Siegfried line. That deployment – perhaps fortunately for those involved – never took place and 'Nellie' was never to dig for victory. The cost of the abandoned project was £8 million: the price, in 1940, of a 35,000-ton battleship. QED.

has, alas, not come down to us. He was told that what Churchill had in mind was a machine or, rather, 200 such machines, which would steal up to the German fortifications in the dead of night, each machine digging *en route* a 7ft (2m) wide and 5ft (1.5m) deep trench behind it from a starting point behind the Allied front line. From this cover, at first light, battalions of infantry, machine-gunners and artillery would emerge to launch a devastating and decisive attack on the concrete fortifications of the Siegfried line.

Goodall was told that £100,000 was to be placed immediately at his disposal. He engaged the services of a young naval architect, with the fitting name of Spanner who, with three naval draughtsmen, was locked into a bedroom suite of the Grand Pump Room Hotel in Bath, and were set to work under conditions of utmost secrecy on their improbable assignment. The project was to have various code-names but the first, and the one which stuck, was 'Naval Land Equipment' (NLE), which became 'Nellie'. Other names used were 'White Rabbit 6' and 'Cultivator 6'. The '6' is said to denote the sixth such war-winning idea to emerge from the fertile mind of the 'former naval person' (i.e. Churchill) during 1939. After a month's work, Spanner had six tentative designs drafted and set out for Lincoln and the offices of Ruston-Bucyrus, the renowned makers of excavators. The chief engineer of that company was, in 1939, Mr Savage. What he thought of the briefing is not known but the specifications that Spanner presented to him were: 'Dimensions 75ft by 10ft, Weight 125 tons, Turning circle 1 mile, Digging rate 100 tons per minute, Speed of Advance ¾ mile per hour, Horse Power from Rolls Royce Merlin engine 1200, Cut 7ft 6in wide by 5ft deep.' Spanner helpfully added that an army patrol could, if required, furnish him with samples of the earth in front of the Siegfried line. Sufficient machines were to be built to create a front 25 miles (40km) wide. The preferred drawing revealed a very slender, low, armoured structure propelled by twin caterpillar tracks some 35ft (10.5m) long, making Nellie the longest tracked vehicle known. At the front was an enormous snowplough-shaped blade, under which the bucket wheel revolved, which cut the trench shaped and widened by the snowplough. The spoil from the excavator buckets was discharged through vents left and right of Nellie and were also formed by the snowplough blade into twin parapets at the top of the trench. The engines – now by Paxman-Ricardo – were inside the hull as were the gearboxes, steering, drive to the tracks and crew. The entire structure was covered by a steel carapace that had to withstand small arms fire and shrapnel.

At this distance of time it seems unbelievable that a nation about to fight for its very existence should seriously have considered such a proposal. But it did. Winston Churchill became Prime Minister, France fell, Dunkirk occurred and the Battle of Britain was fought. The Battle of the Atlantic very nearly brought the country to its knees. In the Far East Singapore fell. The British Army was in retreat on all fronts, and the prospect that any British troops in the foreseeable future would be in a position to hurl as much as a hand-grenade at the Siegfried line was distinctly remote. Nellie proceeded. The work was in the 'Most Secret' category, which made the task of Savage and his staff all the more difficult. All doors, bar one, to the erection shop in Lincoln where Nellie was being built were bricked up. Armed guards kept a twenty-four-hour watch. Suppliers – there were to be 350 of them – were not told what they were supplying for, which caused confusion. The Ministry of Defence refused to sanction the supply of 200 modified Rolls Royce Merlin aero-engines, the design called for, on the not unreasonable grounds that the RAF had a prior claim. By the time the prototype Nellie was ready for field trials, the prospect of using it in the intended role had vanished – if, in fact, it had ever existed. The British Army had not been consulted and the Royal Navy washed its hands of it. One minute sums it all up; it is from the naval officer responsible for Nellie to a War Office liaison officer: '. . . instructions were approved [to build the prototype Nellie] only in the sense that they were approved by the Prime Minister . . .'

The trials of the full-sized prototype (there had been trials of a working model built by Basset-Lowke) took place under the command of Major Whitehouse of the 796 Mechanical Equipment Company of the Royal Engineers, in Clumber Park, Nottinghamshire, the Prime Minister himself being present as was witnessed by contemporary photographs. Surprisingly, Nellie seemed to work rather well, but there were problems: there were clogged tracks, the driver could not see where he was going, the device was underpowered and the engines overheated. Even though Nellie did dig a reasonable trial trench, the question was, at a time when the war was one of mobility and dominated by air power, simply what was the tactical role of this weapon? That question was put in the form of a memo from the production committee. It is a masterly example of Whitehall understatement: 'The tactical problems involved in the use and deployment of these machines appear to this committee to be much greater than the manufacturing problems . . .'

Nellie never dug in anger. The order for production was rescinded on 1 May 1943. It would have been appropriate if it had been exactly one month earlier. Nellie herself was retained to the end of the war on Churchill's orders and was used in trials for digging anti-tank ditches. It had cost the country £8,000,000 (a huge sum in the 1940s). Nellie must hold one world record: the most expensive excavator ever built.

FACT FILE 1

Smallest commercial excavator on market
JCB Mini-excavator 801.4
Loadover height: 7ft 8 in (2.3m)
Maximum reach: 12ft 10in (3.9m)
Overall width: 3ft 3in (1m)
AUW in working order: 1.46 tons
Maximum bucket tear-out: 2,447-ft lb (338kg m)

FACT FILE 2

World's largest dragline excavator
Bucyrus-Erie 4250W
Overall width: 77ft 6in (23.6m)
Depth of digging: 150ft (46m)
AUW in working order: 12,244 tons
Bucket capacity: 220cu yd/168.3cu m (300 tons)

Bovis
Bovis

The Big Lift: Cranes

Give me a lever and somewhere to stand, and I will move the Earth.
(Archimedes)

ARCHIMEDES' PROPOSITION was sound in principle but hardly practical. However, the lever was a major aid used by civil engineers in ancient times when the only practical source of power was human and animal muscle.

It is perfectly feasible to lift a substantial weight using a wooden lever and a number of men, but the height achieved can only be a matter of inches or centimetres. This is enough, it is true, to raise a heavy slab of stone sufficiently to place a roller or sledge beneath it in order to drag it forward, but insufficient for any construction higher than the limited reach of man. Yet we have the evidence of such major feats of civil engineering as the pyramids and Stonehenge. How were they built? The simple answer is that we do not know and probably never will; that, however, has far from discouraged speculation.

These historic structures have been the object of serious investigation, which has appeared on television. Replicas have been constructed using only the methods thought to have been used by the original builders: huge earth or wooden ramps, levers and rollers. It has been shown that a massive stone weighing several tons can be laboriously raised by levering and packing successive baulks of timber beneath it. A replica Stonehenge sarsen stone was raised using a ramp and an A-frame, which is a form of lever that topples the stone into a foundation pit. All the methods used in ancient times relied on the use of unlimited manpower and animals – the only power available.

In the Vatican library there exists a very detailed description of the erecting of an obelisk during the Baroque period. The record reveals the civil-engineering techniques of the time and the great labour and time required before the era of powered cranes. The obelisk had been brought to Rome from Heliopolis in AD 40 by the emperor Caligula. It had stood for centuries in the square behind St Peter's. In 1585, Pope Sixtus V ordered the architect and engineer, Domenico Fontana, to the Vatican Palace to move the 75-ft (22.5-m) 270-ton stone monument to a new location on the frontal approach to St Peter's.

Fontana had at his disposal a massive wooden scaffold or shear legs built on site from timbers 20in (50cm) thick, 40 capstans and levers, ropes and pulleys operated by no

Modern tower cranes working on the Lloyds building in London. These construction cranes are now a common sight in towns and cities as they are essential to the construction of high-rise buildings. All are erected on the site and remain there until the building is completed.

Erecting a 270-ton obelisk in the Vatican in 1585. This etching shows the long ramp with the obelisk lying ready to be raised, dragged upright by the pull of the men and horses, using the massive shear legs as a makeshift crane. (From N. Zabaglia, *Castelli e ponti*, Rome, 1743)

fewer than 800 men and 140 horses. Such was the importance of the event that the Pope ordered that anyone dissenting in anyway 'whatever be their rank or station' would be liable to 'the pain of our displeasure' and, possibly more to the point, a fine of 500 ducats. The whole, successful, operation took months of detailed planning and over two weeks of actual labour to accomplish. An engraving of the event clearly illustrates the mammoth scale of such an undertaking.

No crane then in existence could possibly have been of any use to Fontana in lifting 270 tons and none is listed in the detailed Vatican account. Light, hand-powered cranes have been in use for centuries and were certainly the only type available in Fontana's time. They possibly originated, as did so much else, with the Chinese, who used the spars of their merchant junks as improvised, manually operated jib cranes to raise and slew casks of spices or bales of silk and other relatively light cargo from ship to shore. That is the virtue of the crane as opposed to the fixed shear legs that Fontana used. A crane can raise, slew and lower a load, lifting it from one point and depositing it at another.

Historically, the limiting factors, in terms of lifting heavier weights, in the development of the crane were twofold: the enforced use of wood in the primary construction and the lack of power other than manual operation. Attempts were made in the Renaissance to apply water power as a means of lifting weights. An engraving from Georg Agricola's *De Re Metallica*, published in Basle in 1556, depicts a water-powered machine, but it cannot be called a crane. It was used or, more plausibly, proposed, for raising water from a flooded mine-shaft. The designer had appreciated the need for a form of gearing in that the large (34ft 3in/10.7m) water-wheel has a capstan extended from the axle, giving about a 10:1 reduction. However, the proposal would have been dependent on a constant supply of running water for its operation; hardly a recommendation for the sighting of a mine-shaft.

The 1556 woodcut also tantalizingly hints at an early form of crane. On the left margin can be seen a ladder-like structure scaled in both senses of the word by a man, with a rope and bucket suspended from above. It is likely that at the top of the 'crane' there was a simple manual windlass. Manual windlasses or capstans were used by the builders of the great cathedrals that rose across western Europe from the eleventh century.

Relatively light loads, a small block of dressed stone for example, could be raised by two men using a windlass and a short wooden jib. For heavier loads, a 'great wheel' was built by the carpenters on site and hauled up to the roof trusses of the unfinished building. Two men could stand upright within the wheel, which had a capstan on the axle and, like a hamster in its wheel, they could walk forward to turn the treadmill and raise quite heavy loads. Even greater wheels were built to accommodate horses too. Just why the builders used to haul the great wheels up to the rafters of the buildings instead of placing them on the ground and running a rope up through a pulley is not clear. Perhaps they did. Surviving written accounts of the building techniques of the Middle Ages are sparse.

Above A 16th-century woodcut of Agricola's method of using water power. (Agricola *De re metallica*, Basle 1556)

The crane only became the major means of lifting heavy objects with the arrival of that greatest of Victorian inventions, the steam engine. Steam power, together with the availability of wrought iron and steel for structural strength, meant that some of the largest cranes ever built, or likely to be built, became a reality.

Size is important when considering cranes. At the turn of the century, Britain, then the world's major sea power, wished to remain as such and turned its attention to improving harbours, naval harbours in particular. With ample government contracts forthcoming, very large cantilever cranes were built for the express purpose of constructing breakwaters to extend existing harbours, one of the earliest being Admiralty Harbour, Dover. Between 1898 and 1909 this harbour, extending to 610 acres (244ha), with a breakwater 3,850ft (1,155m) in length, and 50–90ft (15–27m) in depth, was constructed by interlocking 30-

Below A giant Titan steam-driven block-setting crane engaged in the building of a harbour breakwater at Madras, India. Each of the pre-cast concrete blocks interlocked. The crane ran on rails along the top of the newly constructed breakwater as it was built. Such a breakwater could require as many as 64,000 blocks.

ton concrete blocks which were cast on the site. As many as 64,000 blocks were laid, a total of over 1,920,000 tons of masonry. Every single block was transported and laid on the breakwater by Goliath steam-powered cantilever block-setting cranes.

The Goliath cranes at Dover were about 100 tons in weight and were to be dwarfed by a later, and a truly classic, crane – the huge Titan block-setting cranes manufactured by Stothert & Pitt. A Titan that was unloaded weighed 576 tons, and could handle a load of up to 60 tons at a radius of 100ft (30m). They were self-powered and capable of travelling along specially laid 17ft (5m) wide track on 4 bogies, each with 4 wheels. This made a total of no fewer than 16 wheels, 8 of which were power-driven from the main lifting engine by massive bevel gears and drive shafts.

Titans were usually powered by a twin-cylinder steam engine, though electric versions were available when suitable and reliable supplies were on hand. This was not always the case in such places as the harbours at Vera Cruz, Madras and East London in South Africa, which were just three of the many overseas sites where Titans were exported for harbour and breakwater work. The superstructure of a typical Titan was carried on a massive (up to 40ft/12m) roller-track bearing, which allowed the cantilever arm a full 360-degree rotation, one revolution taking three minutes.

Titans were prefabricated and assembled on site. The precise size was variable since each crane was designed for a given site. A Titan had to travel along the breakwater or harbour wall as it was laid, so typically the supporting structure was 30ft (9m) high. The cantilever (up to 200ft/60m long) supported the steam engine, vertical boiler and control cab at one end, together with heavy ballast weights placed beneath the engine cab, all counterbalancing the longer operating arm. Along the cantilever, drawn by control wires, ran a heavy trolley called the Jenny, which carried the 4-in (10-cm) lifting cable running in a multiple-sheaved block. Existing photographs of a Titan show a block with four pulley wheels.

Since multiple sheaves are an important part of any crane's lifting gear it is worth considering their universal use. The technique came from the rigging of sailing ships. It was discovered that by running a rope through double pulleys or tackle, the load a single man could haul was very much increased. In maritime terms this is the 'purchase' which, in effect, offered a form of down-gearing; the haul would take twice as long but require roughly half the manual effort to raise.

The two blocks are defined by sailors as the standing block, i.e. the one which does not move, and the running block, the one which, when applied to a crane, contains the business end with the hook. In simple terms the mechanical advantage, or purchase, is defined by the number of sheaves in the blocks. A two-sheave block would offer a mechanical gain of 4:1 compared with a simple, single-pulley purchase. However, due to friction losses in the bearings of the sheaves, the practical figure would be lower: around 3.57:1, still a considerable advantage. The four-sheave block used in the Titan meant that the load could only be raised and lowered very slowly which, when handling a 40-ton block of concrete, held only in a friction grab, received few complaints from the men beneath who (if they were wise) would keep well clear.

If the Titans are classic cranes, they were also the inspiration for another, though admittedly minor, classic crane – the model of the Titan made from the famous prewar Meccano construction sets. From 1929, when the 'Giant Block-setting Crane' was first announced in the *Meccano Magazine* through to the present day, it has been the biggest

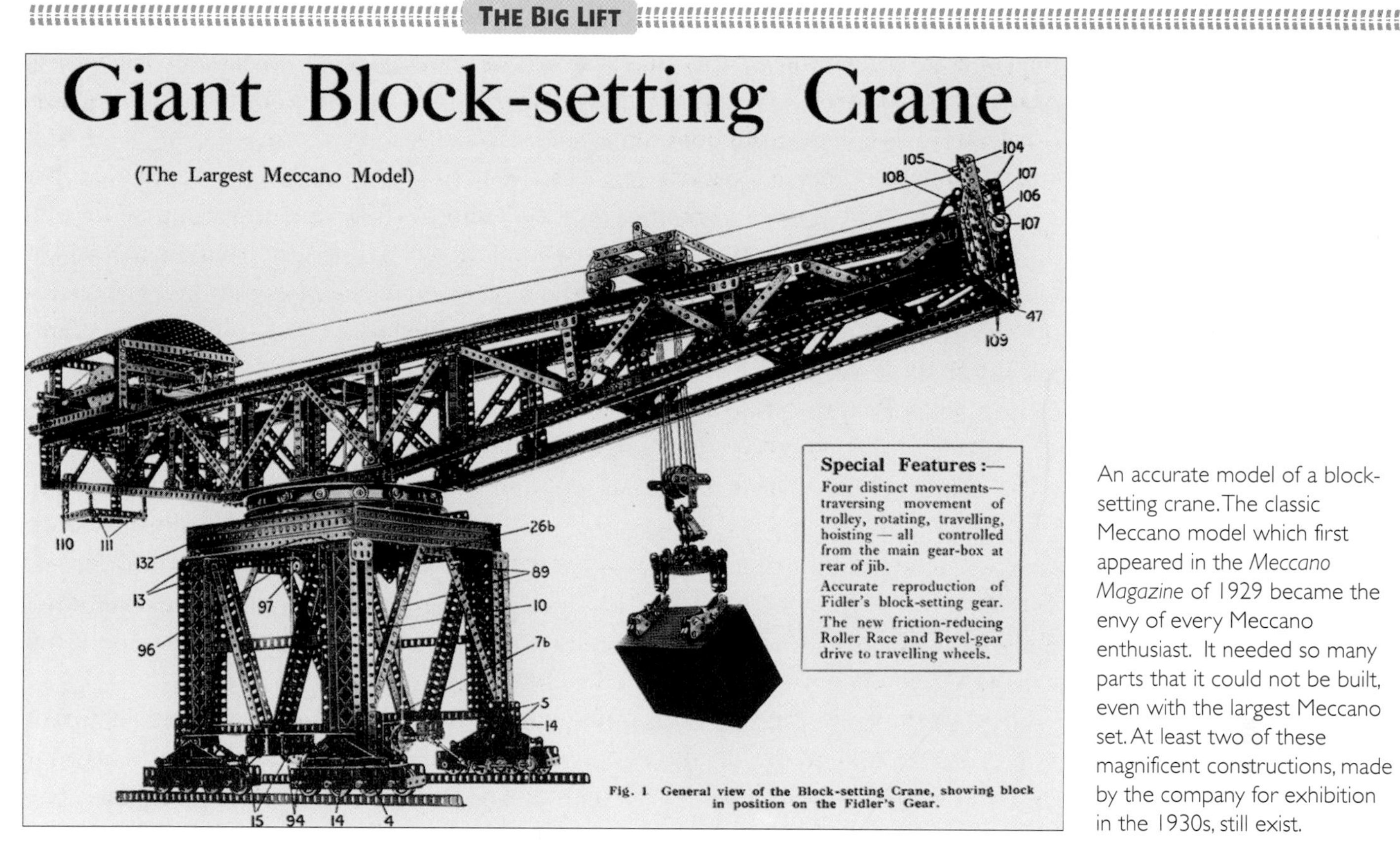

An accurate model of a block-setting crane. The classic Meccano model which first appeared in the *Meccano Magazine* of 1929 became the envy of every Meccano enthusiast. It needed so many parts that it could not be built, even with the largest Meccano set. At least two of these magnificent constructions, made by the company for exhibition in the 1930s, still exist.

model ever offered by the company, and the fascination and despair of Meccano enthusiasts of all ages. Before the Second World War the factory constructed several versions for exhibition and display; the crane also figured prominently for decades in Meccano publicity. At least two of the factory-built cranes survive. It is doubtful if many were built privately; the crane required no fewer than 1,776 listed parts – far more components than were contained in the largest of the prewar Meccano sets. Lesser cranes were built by budding engineers; the Meccano system lent itself admirably to crane construction.

The firm of Stothert & Pitt, the builders of the prototype Titan, were among the first British companies to offer cranes. They were an engineering company of repute, making early steam railway engines, mill engines and a bridge or two. As early as 1850 Stothert had built hand-operated 3-ton quarry cranes. By the 1860s steam-powered cranes were becoming the main business of the company. In 1867 they showed a 6-ton steam-powered crane at the Paris Universal Exhibition at which it was awarded a silver medal. Stothert & Pitt cranes were to win gold medals in London in 1885 and Paris in 1889.

The Goliath and Titan cranes were used mainly for the construction of harbour facilities, and it was in those harbours and docks that the proliferation of cranes took place from the 1890s to the 1960s when containerization was profoundly to alter maritime cargo-handling worldwide. Containerization meant the placing of general cargo into standard containers at source (a works, factory or distribution warehouse) and then transporting the containers by road or rail to dedicated container ports and on to specialized container ships, without any intermediate handling by dockers loading myriad separate items of cargo into the holds of ships with dockside cranes. To the confident Victorian engineers designing and constructing bigger and better docks with steam-powered cranes, containerization and all it foretold lay in the unknown, far distant future.

In 1878 Stothert & Pitt built, to the patented design of William Fairburn, a steam-powered dockside crane with the then unheard-of capacity of 35 tons. The crane, which still stands preserved in perfect working order in the Bristol docks, has a most elegant wrought-iron jib and looks surprisingly up to date despite its age of 120 years. Though now only operated as a tourist attraction in the summer, the crane, which has its base sunk 15ft (4.5m) below the dockside, was fully employed during the Second World War lifting damaged light naval vessels and landing craft out of the water to the quayside for repairs. Andy King, who runs the crane, says of it: 'It was amazing when it was first built; it could lift 35 tons at a time when most could lift only 3 tons. I still think it looks modern with its immensely strong wrought-iron jib. It knocked spots off anything at the rival docks at Cardiff and Newport.'

The classic dockside cranes that could be seen in rows lining every quay from the turn of the century right up to the introduction of containerization in Liverpool, Avonmouth, London and indeed practically every port worldwide were, unlike the Fairburn, mobile. Dockside cranes, of which Stothert & Pitt and Mather & Platt were major manufacturers, were powered by electricity. They ran on rails that traversed the quayside, allowing perhaps up to four individual cranes to work precisely opposite the holds of the ship being loaded or unloaded. Each dockside crane had a capacity of around 25 tons.

Because the cranes had to raise and lower cargo from deep within the ship's holds, they evolved a unique profile: the lower structure comprised four slender lattice girder slim legs some 30ft (9m) long, each with powered wheels guided by the tracks. The slewing part of the crane, the electric lifting and walking motors, the jib and other controls were contained in an extensively glazed cab. The primary jib rose from the sides, and at the apex was another pivoted short jib. The slim jibs of several cranes all in constant independent movement strongly suggested a pride of tall giraffes browsing.

The Fairburn crane at Bristol docks: one of the oldest cranes still in working order in Britain. It was built in 1878 to the designs of William Fairburn by Stothert & Pitt of Bath and is steam-powered. When new it could lift 35 tons, making it the most powerful crane in the Bristol docks. With a base for the fully-slewing mechanism 15ft below ground, it is extremely stable and was working regularly until containerization in the 1950s. Now preserved, the Fairburn is still worked on occasions.

The drivers of these dockside cranes, who incidentally had to stand to operate, were highly skilled. They had to be. Once the hook had disappeared from their sight into the deep hold of the ships, they had to rely on a man who, standing peering into the dark hold of the vessel, relayed the shouted instructions of the dockers below with hand signals to the crane driver 40ft (12m) above the quayside. In a freighter's hold several dockers would be manhandling the cargo and attaching or detaching the loads. Any misjudgement on anyone's part would mean that a serious injury or a fatal accident was inevitable. It was dangerous work; general cargo was stacked in the holds, one on top of another. A cargo, however carefully loaded, might well have shifted during a rough passage. Between the wars, the rules about toxic or dangerous cargo were, by the standards of today, relaxed.

The electric power for the motors came from long cables that snaked across the quay as the cranes 'walked'; each crane had sufficient cable for its limited travel. The cranes worked very quickly: demurrage and missed tides could erode a shipowner's slender profit margins. The cargo, which might comprise sacks of grain, frozen sides of lamb or beef, bales of cotton, and so on, all apparently casually slung with a chain or hooks according to long-established practice, would be unloaded and would land at the feet of gangs of dockers who would carry it away into the warehouses on the quay, the tallymen noting everything in their ledgers.

These long, high Victorian warehouse buildings were romantic places smelling of molasses, grain, spices and brown sugar – redolent of the vast British Empire beyond the seas. Certain cargo, instead of going into the warehouses, would be loaded directly into waiting railway wagons. Railway tracks were always to be seen running along the quays beneath the cranes. Most docks had their own small, standard-gauge steam engines for shunting, and an intricate network of tracks and sidings. The tracks ran both sides of the warehouse. On the side away from the quay, many of the tall Victorian buildings had, on the top floor, a small crane, usually fixed to the warehouse wall and known as a jigger. Each floor of the building had a large doorway, which could be opened for the jigger to load a specific item of cargo from the floor down to a railway wagon or lorry waiting below.

Dockside cranes preserved at the Avonmouth docks, in Bristol. The extensively glazed cabins provided the driver with a clear view of the load he was lifting – essential when working deep in the holds of a ship alongside the dock.

Although dockside cranes performed a major task in loading and discharging cargo the ships, before containerization, had their own cranes in the form of deck derricks. These were steel booms supported by a wire topping lift and pivoted at the derrick heel near the base of the ship's mast or, in the case of some freighters, from two short uprights called Samson-posts. The derricks were either electrically powered from the ship's supply or operated by small steam donkey engines. They were principally used in distant natural harbours that had no docks as such, but ships anchored offshore discharged their cargo directly by derricks into waiting lighters moored alongside the vessel. Interestingly, the term 'derrick' had nothing to do originally with shipping, being the name of a hangman who, in the sixteenth century, was well known for his work at public executions at Tyburn in London (the site is now Marble Arch). Perhaps the name

A dockside scene typical in all British ports until the advent of containerization changed docks for ever. The cranes are loading and unloading ships to and from the quayside. The cranes were the only powered aid; all other cargo handling was done by the dockers.

became associated with the handling of heavy weights, or perhaps a large number of ex-sailors ended up in the hangman's noose.

In January 1965 the funeral of Sir Winston Churchill was televised by the BBC, and there was an unforgettable sequence involving cranes. Sir Winston's coffin was conveyed for part of the procession aboard a Thames boat through the Pool of London; as it did so, the dock cranes lined up on the quayside slowly dipped their tall, slender jibs in silent tribute. There were few, if any, ships at the docks; the cranes were in fact signalling their own farewell. By that time, in 1965, container transport, which had started in 1956 with Malcolm McLean, was soon to end for ever the bustling era of the dockside crane and, indeed, the close-knit working society of dockers, wharfingers, longshoremen, crane-drivers, lightermen, tallymen and the long-established warehouse companies. Going, too, were the imperious liners, the tramp steamers and the rusty freighters with general cargo in their battered holds.

Now, with the millennium within hailing distance, the ships, dockers and cranes are gone. Tallymen and the other tradesmen are now long banished from the major ports, which have become a wilderness of stacked containers and the strange machines that carry them to the waiting ships. Few people seem to be needed in this efficient landscape. Where the dockers and tradesmen once worked, Porsche cars stand parked, their young city-dealer owners living in the warehouses which have been converted into fashionable, luxury flats. The docks themselves – those not yet filled in – have become marinas containing glass-fibre leisure yachts and motor cruisers. Trade is now far down the river, where giant, specialized machines – they cannot be called cranes – lift identical containers and, like a child with building blocks, stack them into and on to the container ships.

Although the major ports of the Pool of London, Liverpool and Southampton have lost their quayside cranes one other port has not: Bristol's Avonmouth docks. The Royal Edward Dock at Avonmouth is one of the few remaining general cargo docks. Although a shadow of its former existence it does exist as a dock with quayside cranes on rails, as all docks in Britain once did. By the dock gates there is a 25-ton crane that dates from 1911, built by William Arrol. It is not a museum piece. It is one of the oldest, if not the oldest, electric crane which is still working commercially.

The quays at Avonmouth have a number of standard, classic dockside cranes. They are relatively new, having been built in the late 1950s and early 1960s, replacing older, now scrapped, steam and electric cranes. Among the new cranes are a Stothert & Pitt Blue Streak. These modern electric cargo cranes are much favoured by the crane-drivers as they allow them to sit down, which was an impossibility with the earlier models. The Royal Edward Dock, though a commercial undertaking, is in a sense an anachronism. In the age of containers it represents a vanished era, but coasters and freighters still moor alongside the old stone quays to load or unload general cargo in the traditional manner by crane.

The old, classic dockside cranes, though mobile, were only so in a very limited sense. They could move along the quayside rails just to position themselves for easy access to the various holds of the ship they were dealing with. Truly mobile cranes, developed following the advent of the railways, were for a long time confined to that industry. These were the 'breakdown' cranes, the term masking their true function, which was to clear the tracks of the wreckage following a railway accident; this was always coyly referred to by the companies as a 'mishap', however serious.

The preserved Arrol crane at Avonmouth. This 25-ton crane was built in 1911. It remains in working order and is by some margin the oldest electric crane still in (occasional) commercial use.

Breakdown cranes were, and are, mounted on standard-gauge railway bogies. In the steam-engine era the breakdown cranes, too, were steam-powered, though not self-propelling. They were drawn by a light engine, the crane engine providing power only for slewing and lifting. The breakdown train consisted of the locomotive, the crane, a number of open 'match trucks' for the wreckage, tool vans with heavy jacks, cutting gear and other tools with, most importantly, baulks of timbers for packing, shoring and various other purposes. There was also an old carriage doing duty as a mess coach for the breakdown crew. The lifting capacity of the cranes was limited to around 25 tons or so by the necessity of conforming to the railway loading gauge, which also required the crane jib to be lowered during transit. It was usually carried resting on a four-wheel, flat bolster truck known as the 'runner'. The base on which the crane operated was restricted by the narrow 4ft 8½in (141.3cm) of the standard railway-track gauge.

One breakdown crane, which has survived from the age of steam, was built in 1907 for the North Eastern Railway. It bears the stock number CME 13 and is preserved, in working order, as part of the National Railway Collection. This crane, despite its early construction, is typical of the type that was in widespread use in the British system until the end of the steam era. Breakdown cranes were fairly standard for, unlike locomotives and rolling-stock, they were not built at Crewe, Derby or Swindon but by specialist firms, the best-known ones being Appleby, Cravens and Cowans-Sheldon.

A standard breakdown crane, such as CME 13, built by Cravens, could clear the tracks of wreckage following a 'mishap', and re-rail the locomotive and any undamaged rolling-stock. Two working together could deal with a more serious accident. Though the track gauge restricted the capacity of the crane when slewed, outriggers in the form of strong girders could be manually extended from the crane and, when suitably packed with timbers and jacks, would stabilize the crane if required.

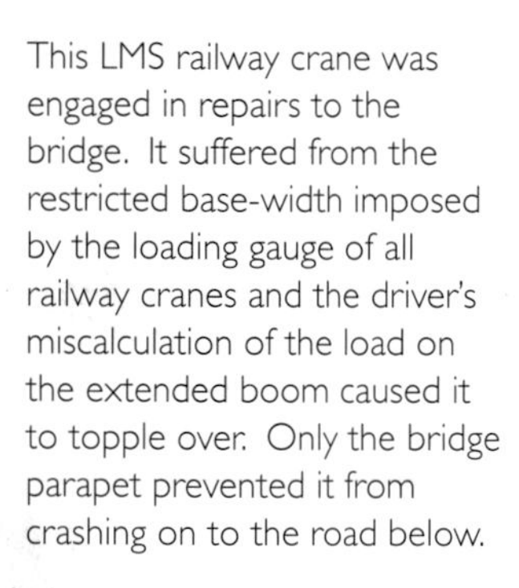

This LMS railway crane was engaged in repairs to the bridge. It suffered from the restricted base-width imposed by the loading gauge of all railway cranes and the driver's miscalculation of the load on the extended boom caused it to topple over. Only the bridge parapet prevented it from crashing on to the road below.

Steam-powered breakdown cranes were still being built for export to the railways of India as late as 1960. In Britain, since the ending of the steam era, some of the old breakdown cranes have been converted to diesel-electric operation. In 1960 one of the original builders of breakdown cranes, Cowans-Sheldon of Carlisle, built a massive diesel 'wrecking' crane with a capacity of no less than 250 tons for a railway company in Canada. No railway crane of that size could conform to the British loading gauge. Even the American loading gauge proved a limiting factor in one notable accident in the 1940s, which involved the overturning of one of the mighty, 300-ton 'Big Boy' steam engines (the largest ever built). No existing US railway wrecking crane could lift it. The locomotive had to remain where it lay for several weeks until a purpose-built crane of sufficient capacity was designed, prefabricated and erected on site to lift and re-rail the stricken giant.

In Britain, with the demise of the 140-ton steam locomotive, modern diesel-electric breakdown cranes with a modest 25-ton capacity can deal with any 'mishap' involving contemporary electric locos. Breakdown cranes apart, all the pre-British rail companies had other, lighter mobile railway cranes in their stocklists. The civil engineering department would have at their disposal a number of medium-capacity – 10 tons or so – steam-powered cranes mounted on four-wheel trucks for such duties as lifting replacement signal posts, bridge girders, prefabricated sections of track and other duties. It has to be said that it was by no means unknown for these cranes to be overloaded and topple over when slewed; a consequence of the restricted narrow base.

At the lowest level of railway cranes were the universal 'goods' cranes, once to be found on every station, even on the most humble branch line. These cranes were located, on a fixed base, in the goods shed or at the far end of the platform. They were hand-operated by the staff slowly – very slowly – turning a handle connected to cast-iron gears and so drawing the single lifting cable. These universal, basic cranes, dating from the turn of the century, had a cast-iron jib, also hand-slewed. Their function was to unload heavy items, up to about 1 ton: a new Fordson tractor or Ransomes plough, sacks of feedstuff or cases of furniture. These were the warp and weft of the steam railway's goods system, all brought in wagons and vans, and shunted into the siding by the daily (or weekly) 'pick-up' goods train. The small railway cranes are classics in the sense that the steam-railway network they served, and were part of, is now itself classic.

Railway cranes apart, a degree of mobility was possible in the general-purpose cranes which worked in quarries, building sites and factory yards. These cranes, of varying capacities, had limited movement. Beecham's chemical factory used a typical prewar crane, a 1938 Rapier, which moved, self-powered, on solid tyres. Such a crane could only be expected to work on a solid, level, paved yard. Building sites certainly could not offer that and, in the 1930s, cranes were evolved that moved on crawler tracks, first developed by the American Caterpillar Company. Although the ability of cranes to move within a limited radius of the site was essential, if they had to move to a new site any distance away they had to be loaded, and partly dismantled in some cases, on to heavy trailers and towed to the new location.

The first truly mobile crane, a road-going version of the railway cranes, is credited to a British company: Coles Cranes. Henry Coles, born in 1847, worked as a young man with the London crane-makers, Appleby Brothers. He left the company with his three younger brothers, all of whom had worked at Appleby Brothers, to form their own London-based crane and engineering firm, Henry J. Coles, in Southwark.

Coles, a leading name in cranes for over a century are still a major supplier of modern cranes. This etching shows what is considered to be their first. One of six steam 'railway', fully-slewing, self-propelled cranes ordered c.1890 to work in the Glasgow Gas Works. The one depicted is fitted with a grab, possibly loading coal from the railway wagon in the foreground to a hopper for the gas plant.

Profiting from the experience gained working for the Appleby company, the firm was a success almost from the start and was to become the foundation of the company which, via take-overs, remains a market leader to this day. By 1890, the cranes that were being offered by Coles were all railway-mounted and steam-powered, and were selling well worldwide. Always an innovating company, Coles invented, patented and marketed a self-releasing dredging grab which could be fitted to an existing crane. Cole's 'Single-chain Grab Dredger' was featured at the International Inventors Exhibition in 1884.

Six Coles steam cranes, railway-mounted, fully slewing and fitted with the patent grabs, were sold to the Glasgow Gas Works in 1890; they were used to lift and move coal and coke. Coles also listed a 'Goliath' crane (Goliath was a popular name for large cranes; this one had no connection with the Stothert & Pitt block-setting cranes). Coles's Goliath was a gantry crane. Only one seems to have been sold, however; at £550, a fortune at the time, there were few takers. Coles also used the availability of the London hydraulic ring main, used principally to power city lifts, including many of those in use at London Underground stations, and to power three fixed-base cranes at the Woolwich arsenal. Clearly, the use of steam-driven cranes would have been most unwise in the handling of the high explosives that were the stock-in-trade at the arsenal. Coles was not the first company to build hydraulic cranes; Newcastle docks had some as early as 1874. Coles was, on the other hand, the first company – in 1897 – to offer a rail-mounted, mobile, hydraulic crane. Just how the hydraulics were applied is not clear, possibly by an armoured hose, which must have restricted mobility. By the mid-1890s, Henry J. Coles Ltd, which had outgrown its Sumner Street premises in Southwark, moved to Derby, although they called their new factory 'The London Crane Works'. They were to remain in Derby for the next fifty years.

The company marked the years with some significant milestones. In 1907 a 40-ton steam crane on a standard-gauge railway base was the heaviest crane the firm had built to that date, but perhaps the most significant development was made in 1922. This was the first road-mobile crane. It was the idea of Arnold Hallsworth, who joined the firm in 1918 as an indentured apprentice but who would, in later years, become the managing director of the company. Hallsworth had proposed building a general-purpose 2-ton crane on a Tilling-Stevens petrol-electric bus chassis. The Tilling-Stevens chassis used a petrol engine to generate electricity, which was used to power the traction motor, thereby dispensing with clutch and gearbox. The electric power was equally available to power the crane drive as well. As is often the case with innovative designs, the trade response to the mobile crane was lukewarm; the early postwar years were a period of uncertainty and depression. There was also a certain Luddite attitude to the idea of a motorized mobile crane. A few were sold to Japan and India for harbour duties but the mobile crane was hardly a commercial success.

The original Coles company, a family-owned business, changed hands in 1926, though it retained the Coles name, and took on the 27-year-old Arnold Hallsworth as general manager. The Depression of the 1930s hit the firm hard and orders were few and far between. By 1936, the workforce was down to fifty people, only three or four cranes being ordered a year. Among the few cranes made was a mobile yard crane powered by a 25-hp diesel engine. This was Coles's first diesel crane. Few were to be ordered. The survival of the company depended on other engineering work and the supply of spares for cranes already in service.

Despite the knowledge that bankruptcy and ruin faced the tottering company, Arnold Hallsworth, with the optimism of youth, tendered for an Air Ministry proposal that called for a 2-ton, fully slewing self-propelled crane, mounted on road wheels, to be known as an Electric Mobile Airfield (EMA) crane. An order for 120 cranes was awarded following a successful tender. Hallsworth drew the outline of a prototype, which the Air Ministry accepted, along with an order for two prototype Coles EMA cranes for evaluation by the RAF.

Coles's proposal had been little more than a hasty, paper outline; no details had been worked out or costed, and no engineering drawings made. By subcontracting and working around the clock, the first prototype Coles EMA crane was built on a Morris Commercial chassis and delivered for evaluation. It more than fulfilled the ministry's requirements: eighty-two were ordered, the largest single order ever made by a British crane-maker (the original order for 120 EMAs had, earlier, been reduced). The date of the ministry contract was 13 July 1937, and this was the date on which the Coles company was saved. The concept of that EMA crane, a diesel engine driving a generator with electrical control of the power drive, was to remain the basis of the company's cranes for the next thirty years.

Coles's EMA cranes were to be seen on every RAF airfield, as well as Allied ones, during and after the Second World War. They lifted crashed aircraft, changed engines

The RAF had a large number of Coles cranes of various loadings. Salvaging crashed aircraft was a common wartime task, though many were employed, as is this one, in effecting an engine change – in this case on a Halifax bomber c.1944.

and did all the many tasks required on operational bases. Airfields apart, there is a photograph of Coles's mobile cranes in use by US Army engineers in action during the Rhône crossing of 1944. Sixty EMA cranes, on Thornycroft chassis, were built at Derby, and another 120 in a new Coles plant located at Sunderland, following a 1942 amalgamation with Steel of Sunderland (the company name Coles Cranes was retained because a proposed change to Steel Cranes was summarily rejected by the board).

In 1945 the war ended. The postwar years were to be ones of reconstruction, and cranes of all types would be in steady, growing demand. The snag for Coles was that their wartime contracts had been EMA cranes built in seven versions, all to strictly military requirements, which meant, in practice, a limit of 6 tons' capacity.

Since Coles had been the main military EMA contractors, their competitors, who had built larger industrial wartime cranes, were now about to corner the postwar commercial market. This was at a time when military orders had abruptly ended and the market for mobile cranes was flooded by the sale, at knock-down prices, of war-surplus EMA stock.

Undaunted, the Coles company enlarged the works and began an aggressive sales campaign backed by an excellent design team who produced a range of new, high-capacity civilian mobile cranes to meet the growing need for reconstruction following the end of the war.

Mobile cranes far exceeding the EMA 6-ton wartime limit were built, and 20-ton-capacity mobile diesel-electric cranes appeared in the 1950s. There were technical developments, too. In 1959 Coles Cranes merged with two minor crane companies, R. H. Neal and F. Taylor and Sons, which had been developing the modern concept of hydraulic, modular, telescopic cranes.

The result was mobile cranes with hydraulic telescopic booms and diesel-hydraulic transmission. The telescopic jibs, unlike the traditional lattice jib, were built from steel-box sections that telescoped into each other, and were extended or retracted by inbuilt hydraulic jacks under the fingertip control of the driver from his cab. By employing this method, the crane jib could be extended by a factor of four, enabling a very tall crane to be deployed on site. When the work ended the jib could retract in minutes to lie within the wheelbase of the vehicle like a fire engine's ladder. Sold initially under the name of Coles-Hydra ('Hydra' had been a Taylor trade name), a very large range of cranes was available, for sale or contract hire. In transit, the Coles mobile cranes were legally roadworthy.

The telescopic cranes were produced at the same time as the traditional lattice mobile cranes but the capacity rose to surpass any previous mobile cranes. In 1954 the 41-ton Coles Colossus was announced. This was, at that time, the largest mobile crane in the world, and contributed to the speedy construction of the British motorways. Then in 1963 the first 100-ton mobile crane, the Coles Centurion, set another world record: it was the first mobile crane able to lift 100 tons. By 1971 that figure was to be doubled with the 200-ton Coles Colossus. These very large cranes were transportable rather than truly mobile like the telescopic cranes. The largest would arrive on site as a kit of parts on several lorries and a skilled team of erectors would require thirty-six hours, using a telescopic crane, to assemble them fully.

In contrast, the highly mobile, telescopic, smaller cranes are to be seen today in works yards, on building sites and anywhere that a mobile crane is needed. Many are on contract hire, a principal hirer being Sparrow Crane Hire, started in 1945 by the Sparrow brothers with ex-War Department (WD) cranes. Today the company is international and the leading source of crane hire up to a 500-ton capacity. On hire, or owned outright, these cranes drive to the site, at a maximum speed of 40mph (64km/h), extend the hydraulic outriggers and the telescopic jibs, and are at work in a very short time. They are classics and are now offered by many crane-makers in many countries.

Other contenders for classic status must be the giant lattice-tower cantilever cranes that dominate the skylines of our cities and towns during the construction of high-rise buildings. A typical maker is Beck & Pollitzer. These very tall cranes are erected on site with the aid of mobile telescopics. The higher these cranes are, the less they can safely lift. However, as their loads are mainly building materials and not unduly heavy, they are therefore lightly built and have to take into account the wind strength as the light, long, cantilever jibs tend to 'weather-cock'. A standard fitting on the top of the cantilever arm is an anemometer to advise the driver of the wind strength. The stabilizing outriggers at the base of the tower can extend across the entire site. It might be whispered that it is not entirely unknown for these daring cranes to topple over. Are they classics of the future? Perhaps, but they are ephemeral – a massive Meccano set to build a crane to suit a given site which is then dismantled and put back into the box. True, they are among the most impressive cranes ever built, with the possible exception of those magnificent Titan block-setting cranes of a hundred years ago. For many these remain a truly classic plant machine.

Every age leaves its monuments, totems and artefacts for the following generation to make of them what it may. Most of the constructions that have survived from antiquity are either religious or ritual: the pyramids, Stonehenge, Inca temples and the cathedrals of the Middle Ages. Even the magnificent docks and railway stations built by the Victorians were constructed with an almost theological zeal and conviction. All are substantial structures built by the hand of man. A cathedral that took a hundred years to construct will not be lightly demolished to make way for a replacement. Today we have no such inhibitions. We have the means, with modern plant and materials, to build a Canary Wharf in a year or so; when it has served its purpose it will make way for another more apposite structure. The classic plant we have been discussing is, it should be remembered, only a means to an end, not an end in itself. It is up to future generations to ensure that computerized mechanization does not simply degenerate into 'improved means to unimproved ends'.

Canary Wharf in London is an icon of commercial building at the end of the twentieth century. The methods used in the construction of such properties depends on the very wide range of modern cranes. Here we see a virtual cross-section of the types available, from the small mobile telescopics to the high tower cranes which have become a feature of our city skylines.

Selected Bibliography

Nicolas Bentley, *The Shell Book of Motoring Humour* (Michael Joseph, London, 1976)

David E. Evans, *Lister's: The First Hundred Years* (Alan Sutton, Dunstable, 1979)

George Ewart Evans, *The Horse in the Furrow* (Faber & Faber, London, 1967)

Roger A. Freeman, *Airfields of the Eighth* (Battle of Britain Prints International, London, 1978)

David Gladwin, *Steam on the Road* (Batsford, London, 1988)

Michael Hancock, *JCB: The First Fifty Years* (Special Event Books, Horsham, Sussex, 1995)

David E. Johnston, *An Illustrated History of Roman Roads in Britain* (Spurbooks, London, 1979)

Peter Kemp, *The Oxford Companion to Ships and the Sea* (Paladin Books, Granada, 1979)

Friedrich Klemm, *A History of Western Technology*, 3rd edn (MIT Press, Cambridge, Mass., USA, 1970)

Charles Knightly, *Country Voices* (Thames & Hudson, London, 1984)

John Mortimer, *The Engineer: Highlights of 120 Years* (Miller Freeman, London, 1976)

The Motor Car, Motor Cycle & Commercial Motor Index (Fletcher & Sons, Norwich, 1929)

Ivo Peters, *Somewhere Along the Line: Fifty Years Love of Trains* (First published by Oxford Publishing Company, Oxford, 1976. Haynes Publications Inc, Yeovil, Somerset, 1988)

Paul Redmayne, *Transport by Land* (John Murray, London, 1948)

L.T.C. Rolt, *Lansing Bagnall: The First Twenty-One Years at Basingstoke* (Sir Joseph Causton & Sons Ltd, London, 1970)

Quentin Seddon, *The Silent Revolution* (BBC Books, London, 1989)

Michael Williams, *Massey-Ferguson Tractors* (Blandford Press, Dorset, 1987)

Geoff Wright, *The Meccano Super Models*, 3rd edn (New Cavendish Books, 1989)

Index

*(Numbers in **bold** refer to illustrations)*

Picture Acknowledgements
Andrew Morland: 6; Rural History Centre, University of Reading: 9, 12, 56, 59, 62, 65, 67(bottom), 68; Brian Johnson: 11, 14, 67 (top); Lee Irvine of the Dan Albone Archive: 13, 15; Ford Motor Company: 20; J. H. Appleyard: 22-23, 26, 27; AGCO Ltd: 29, 32, 70, 73, 79(top, middle & bottom); JCB Ltd: 31, 101, 117; Brian Bell/Farming Press Ltd: 73(bottom); Ransomes Sims & Jefferies Ltd: 78; Tony Stone Images: 34 (Michael Busselle), 50 (Oliver Benn), 54 (James Balog), 74-75 (Zane Williams), 82 (Bruce Forster), 94-95 (Charles Thatcher); Topham: 36, 39, 44, 49, 66, 84, 130; The Road Roller Association Archive Collection: 40, 41, 42, 43; Simon Mein: 46; Stuart Sadd: 131; Roger Savage: 111; Uden Associates: 81 (Anna Whelpdale); Lister-Petter Limited: 87, 89; Lansing Linde Ltd: 85, 86, 91, 93, 96, 97, 99; Lex Service PLC: 100; Bucyrus-Europe Ltd: 102, 107, 115; M.R.P Photography: 114; Kruppfördertechnik GmbH: 118-119; Imperial War Museum: 120, 136; Building Magazine: 122 (Tony Weller); Builder Group: 138-139 (Yoke Matze); MW Models: 125, 127; Richard Hall: 128, 129; The Ivo Peters Collection: 132; Grove Crane: 134